THE GUITAR

A Module on Wave Motion and Sound

FVCC

A. A. Strassenburg, State University of New York at Stony Brook
Bill G. Aldridge, Florissant Valley Community College
Gary S. Waldman, Florissant Valley Community College

Bill G. Aldridge, Project Director

MCGRAW-HILL BOOK COMPANY

NEW YORK
ST. LOUIS
DALLAS
SAN FRANCISCO
AUCKLAND
DÜSSELDORF
JOHANNESBURG
KUALA LUMPUR
LONDON
MEXICO
MONTREAL
NEW DELHI
PANAMA
PARIS
SÃO PAULO
SINGAPORE
SYDNEY
TOKYO
TORONTO

The Physics of Technology modules were produced by the Tech Physics Project, which was funded by grants from the National Science Foundation. The work was coordinated by the American Institute of Physics. In the early planning stages, the Tech Physics Project received a small grant for exploratory work from the Exxon Educational Foundation.

The modules were coordinated, edited, and printed copy produced by the staff at Indiana State University at Terre Haute. The staff involved in the project included:

Philip DiLavore . . . . Editor
Julius Sigler . . . . Rewrite Editor
Mary Lu McFall . . . . Copy and Layout Editor
B. W. Barricklow . . . . Illustrator
Stacy Garrett . . . . Compositor
Elsie Green . . . . Compositor
Donald Emmons . . . . Technical Proofreader

In the early days of the Tech Physics Project A. A. Strassenburg, then Director of the AIP Office of Education, coordinated the module quality-control and advisory functions of the National Steering Committee. In 1972 Philip DiLavore became Project Coordinator and also assumed the responsibilities of editing and producing the final page copy for the modules.

The National Steering Committee appointed by the American Institute of Physics has played an important role in the development and review of these modules. Members of this committee are:

J. David Gavenda, Chairman, University of Texas, Austin
D. Murray Alexander, DeAnza College
Lewis Fibel, Virginia Polytechnic Institute & State University
Kenneth Ford, University of Massachusetts, Boston
James Heinselman, Los Angeles City College
Alan Holden, Bell Telephone Labs
George Kesler, Engineering Consultant
Theodore Pohrte, Dallas County Community College District
Charles Shoup, Cabot Corporation
Louis Wertman, New York City Community College

This module was produced through the efforts of a number of people in addition to the principal authors. Laboratory experiments were developed with the assistance of Donald Mowery and Arthur Noxon. Illustrations were prepared by Robert Day, Daniel Rothwell, and Susan Snider. John Yoder, III, coordinated art work, and he also prepared the Instructor's Manual. Reviews were provided by members of the Physics of Technology Steering Committee and participants in the NSF Chautauqua Program. Finally, Wendy Chytka typed the various module drafts, coordinated preliminary publication efforts, and acted as a liaison person with the Terre Haute Production Center where final copy was being produced. To all of these persons, we are indebted.

The Guitar

Copyright © 1975 by Florissant Valley Community College. All rights reserved. Printed in the United States of America. No part of this publication may be reproduced, stored in a retrieval system, or transmitted, in any form or by any means, electronic, mechanical, photocopying, recording, or otherwise, without the prior written permission of the publisher.

Except for the rights to material reserved by others, the publisher and copyright owner hereby grant permission to domestic persons of the United States and Canada for use of this work without charge in the English language in the United States and Canada after January 1, 1982. For conditions of use and permission to use materials contained herein for foreign publication or publications in other than the English language, apply to the American Institute of Physics, 335 East 45th Street, New York, N.Y. 10017.

ISBN 0-07-001716-6

1 2 3 4 5 6 7 8 9 0 EBEB 7 8 3 2 1 0 9 8 7 6 5

TABLE OF CONTENTS

	Page
Introduction	1
Goals for Section A	1
Section A	2
What Guitars Do	2
Experiment A-1. The Guitar and Its Sounds	3
Summary of the Results of Experiment A-1	8
Vibrations	9
Transverse and Longitudinal Pulses	10
Vibrational Coupling	12
Loudness and Amplitude	12
Frequency and Pitch	13
Experiment A-2. Harmonics and Modes of Oscillation	15
Summary of the Results of Experiment A-2	18
Node to Node Distances and	
Frequencies of Oscillation	20
The Quality of Sound and Patterns of Vibration	20
The Shape of the Sound Board and the	
Location of the Bridge	21
Summary	21
Goals for Section B	23
Section B	24
Qualitative Observations Suggest Quantitative Experiments	24
Experiment B-1. The String Equation	25
Discussion of Experiment B-1	29
Loudness, Amplitude, and Frequency	30
Frequency Response of the Guitar	31
Experiment B-2. Loudness and Resonance	32
Discussion of Experiment B-2	36
Sound Intensity and Decibels	36
Hearing Response	41
Intensity Level in Phons	43
Summary	44
Goals for Section C	46
Section C	48
Traveling Waves on a String	48
Experiment C-1. Transverse Pulses on a Spring	49
Discussion of Experiment C-1	51
Traveling Waves on the Sound Board	58
Mixtures of Harmonics	58
Longitudinal Sound Waves in Air	61
Harmony and Musical Scales	62

Experiment C-2. Guitar Scales 67
Summary 69

Work Sheets
   Experiment A-1 71
   Experiment A-2 73
   Experiment B-1 75
   Experiment B-2 77
   Experiment C-1 79
   Experiment C-2 81

The Guitar

INTRODUCTION

In this module you will use a guitar to learn about vibrations of objects, the nature of sound, principles of wave motion, and the physical basis of music. Many of the concepts and principles presented also apply in some way to other musical instruments and to non-musical sounds. The concepts and principles can even be used to describe completely different things, such as water waves and light.

There are no special prerequisites for this module. It is assumed that you know how to measure length and mass in the metric system of units, given the proper equipment.

GOALS FOR SECTION A

The following goals state what you should be able to do after you have completed this section of the module. These goals must be studied carefully as you proceed through the module and as you prepare for the post-test. The example which follows each goal is a test item which fits the goal. When you can correctly respond to any item like the one given, you will know that you have met that goal. Answers to the items appear immediately following these goals.

1. *Goal:* Understand how the details of guitar construction and how it is played affect its loudness, pitch, and quality.

   Item: Plucking the lowest string of a guitar six or seven inches from the bridge produces a sound of a certain pitch and quality. Using the same guitar, how could you produce a sound with the same pitch but a more hollow quality?

2. *Goal:* Understand the concept of elasticity and how the elasticity of a material affects its vibrations.

   Item: Arrange the following objects in order of increasing elasticity: a strip of paper, a wooden tongue depressor, a thin, flat strip of steel.

3. *Goal:* Understand what is meant by the fundamental and its harmonics, and how these can be produced in a string fixed at both ends.

   Item: How can you produce a guitar string vibration consisting primarily of the fourth harmonic? Draw the pattern of vibration for the fourth harmonic.

4. *Goal:* Know how the pitch and quality of sound are related to string vibrations.

   Item: A guitar string is plucked in a normal way and produces sound with a certain pitch. How can you raise the pitch by one octave without plucking the string again?

Answers to Items Accompanying Previous Goals

1. Pluck the same string exactly at its midpoint.

2. Paper, wood, steel.

3. Pluck the string at the normal position, then touch it lightly at a point one-fourth of its length from either end.

4. Touch it lightly at its midpoint.

SECTION A

WHAT GUITARS DO

Guitars are used to change vibrations of a taut string into sound. Ancient Egyptian and Hittite carvings show guitar-type instruments being made and played as far back as 3000 years ago. One might imagine a man toying with the variations of his bow-string "twang" as the beginning of stringed instruments. By a combination of circumstance and intuition, craftsmen and players over generations have developed the sound board, the fingerboard, more strings, a body, and sound holes in a variety of stringed instruments. Our word "guitar" is much like the old Spanish word *guitarra*, which possibly was derived from the 2500-year-old Sanskrit *chhatur-tar*, meaning "four strings."

The guitar of today is similar to the old Spanish guitar with six strings and an "hourglass" body. This type of guitar, shown in Figure 1, is one of the most popular of instruments. It is estimated that in the U.S. alone, there are 15 to 20 million guitar owners.

To become more familiar with this instrument, do Experiment A-1.

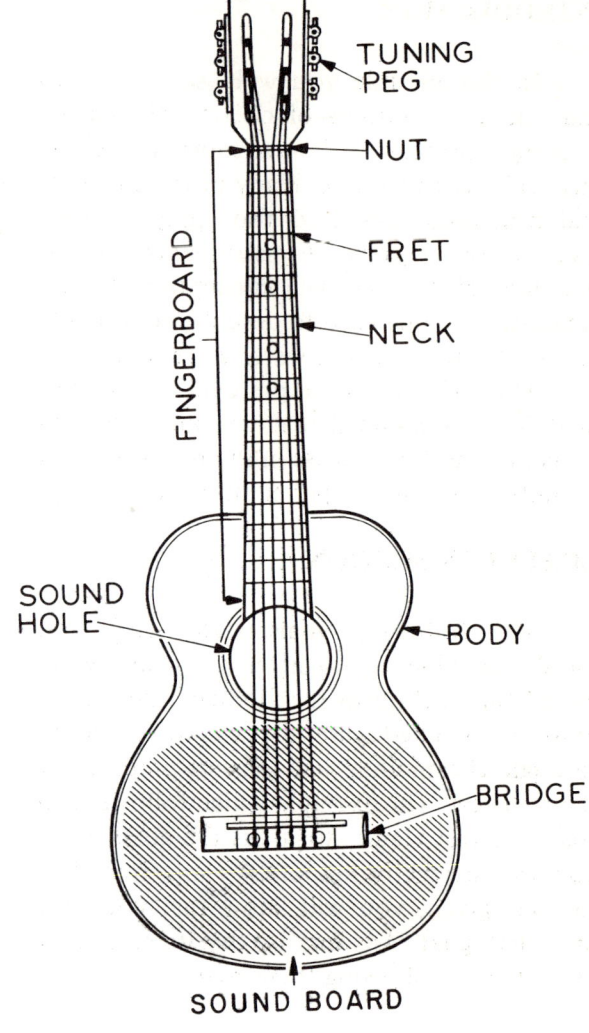

Figure 1.

EXPERIMENT A-1. The Guitar and Its Sounds

This experiment requires the use of a tuned guitar, a guitar pitch pipe, a tin-can-and-wire telephone, and a long brass spring.

Directions to guide your initial study of the guitar are contained in the following paragraphs. Each time you come to a numbered question, write an answer based on your observations on the work sheets provided at the end of the module. (The guitar strings are numbered beginning with the smallest diameter, which is the first or high-E string, and going to the sixth string. The common tuning of the guitar, starting from the first string, is E′, B, G, D, AA, EE.)

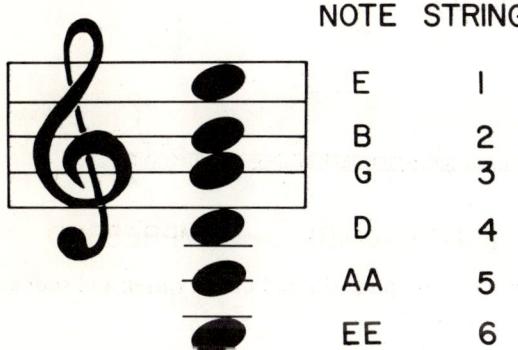

Figure 2. Tuning of the guitar.

Remove the largest diameter string (the low-E or sixth string) from your guitar. Stretch it tightly across your thumbs as shown in Figure 3 and pluck it with your little finger. Notice that the "twang" is not very loud.

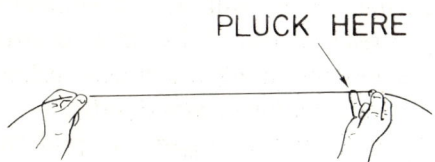

Figure 3.

You might think that the sound hole of a guitar picks up this quiet sound and amplifies it. You can check this assumption by plucking the taut string while holding it just over the sound hole. Be careful not to touch the guitar.

1. Is the sound louder? Does it sound different? If so, how?

As you pluck the tightly held string, feel the vibrations of your thumbs and hands. The vibrating string causes a twang, but a sound also emanates from your vibrating thumb. With your thumb about a centimeter away from your ear, listen to the vibration. Compare the sound of the vibrating string and the sound of your vibrating thumb. By lightly touching your thumb to your ear you can hear its vibrations more loudly.

2. Describe the sounds caused by the vibrating string and your vibrating thumb. How do these sounds differ?

You have just seen that a vibrating string creates sound in two ways: 1. directly, which is the "twang" of the string, and 2. indirectly, by vibrating whatever is connected to its ends. Either way, a vibrating object causes sound. Walk around the lab and look for a variety of objects with flat surfaces such as a table, the floor, a wall, a window, a door, a desk, the chalkboard, a cardboard box, etc. Hold the string taut and press it under one thumb against such a surface, as in Figure 4. Pluck the string. Do this for several surfaces.

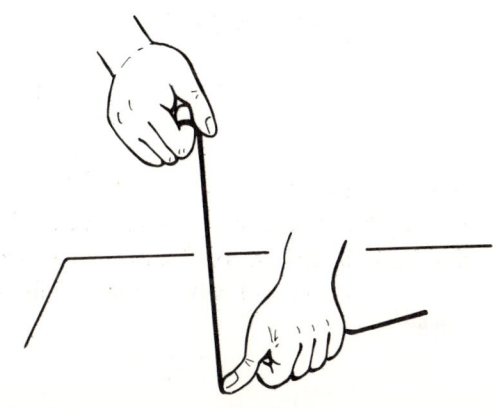

Figure 4.

3. What kind of surface best "sounds out" the vibrations you feel in your hand?

Using the same procedure, sound out different places on the guitar.

4. Which spot on the guitar do you find is best for producing the loudest sounds?

Sound coming from a vibrating surface connected to a string may remind you of the string-and-tin-can telephone which children often make. It consists of a wire stretched tightly between the bottoms of two tin cans. Each can has an open end into which you can speak and listen.

You are provided with such a device. Clamp one tin can to a table as shown in Figure 5 and position a transistor radio so that its speaker faces into the can. Turn on the radio and stretch the wire tight to hear the sound in the other can. First touch the bottom of the other can, then touch the wire.

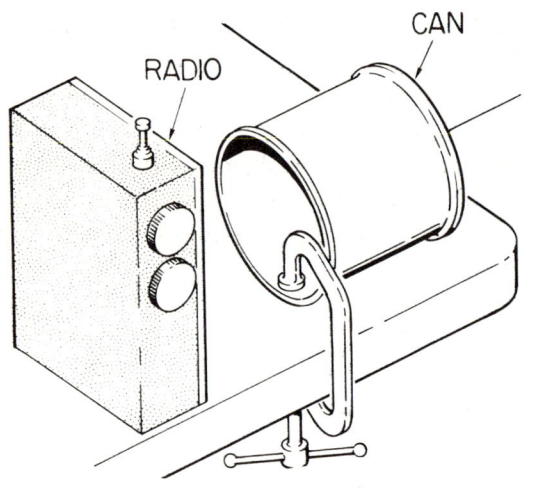

Figure 5.

5. What is the bottom of the other can doing? What happens to the sound when you touch the second can? What happens to the sound when you touch the wire?

Remove the can from the "receiving" end of the wire. To hold the wire taut, take a small length of it near its free end and loop it around your thumb; then press your thumb (with the wire wrapped around it) to your ear. While the wire is pressed against your ear, touch the wire with your free hand.

6. Can you hear the vibrations? Does touching the wire stop the sound? *Can* you stop the sound by holding the wire still? Can you then draw any conclusions about the way in which the wire must be vibrating?

To demonstrate in slow motion two possible types of wire vibrations, we can use a long spring which has been stretched out. (See Figure 6.)

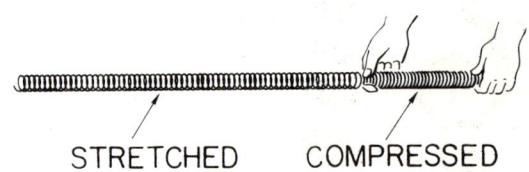

STRETCHED   COMPRESSED

Figure 6. Compress the end of the spring and suddenly release it.

Place the spring along the floor with one end held fixed. Stretch the spring to two or three times its original length, holding the free end in your hand. The spring will now behave much like a taut wire. As indicated in Figure 6, "bunch up" the last few inches of the spring, then let go suddenly. (Do not let go of the end of the spring.) Watch the motion of the spring. You will see a bunched up (*compressed*) section between two stretched sections moving along the coiled spring. This action is called a *longitudinal pulse*.

As indicated in Figure 7, pull the last few inches of the spring to one side and suddenly let go. The resulting action is called a *transverse pulse*.

Watch the motion of the pulse as it

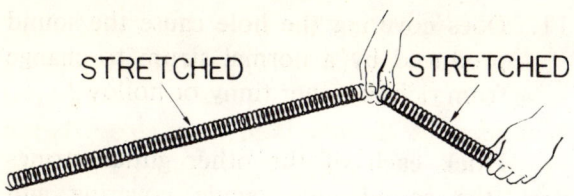

Figure 7. Pull the spring to one side near the end and suddenly release it.

moves along the spring, hits the fixed end, and comes back to your hand.

Repeat this procedure a few times for both types of pulses, paying attention to the feeling in your hand when the pulse comes back and hits it. When a pulse hits the fixed end of the spring and comes back, we say that the pulse is *reflected* at that end. You can detect the reflected pulse best if you do not let your hand touch the floor. To try a combination pulse, compress the last few inches of the spring and simultaneously draw the spring to one side, as shown in Figure 8, then let go.

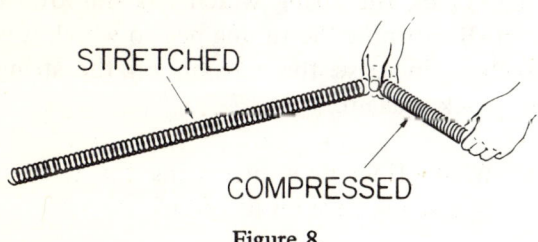

Figure 8.

You have seen (and felt) that the ends of a taut wire can vibrate with a side-to-side motion like that of a transverse pulse. The vibration of the bottom of the tin can implies that a taut wire also vibrates with a to-and-fro motion like that of the longitudinal pulse on the spring. You also found that the guitar bridge, where the vibrating strings are connected, is the best place on the guitar to sound out transverse vibrations.

Re-install the sixth string on the guitar. Following the instructions provided with your guitar, *tune* the string. Then raise the sixth string above the others by sliding a pencil under it, but over the others somewhere near the nut. At or near its midpoint, carefully pluck the string so that transverse vibrations are moving up and down (perpendicular to the sound board). Pluck the string by holding it between your thumb and forefinger, pulling it up, and letting it go. You can determine if the string is vibrating correctly by sighting with your eye. Compare the loudness of the sound you hear to the loudness heard when the string is plucked so that its transverse vibrations are parallel to the sound board. Repeat this observation several times to be sure your results are dependable.

7. Which direction of vibration produces the louder sound? What can you conclude about the way vibrations are transferred to the guitar bridge and sound board? For either direction of vibration, how would you pluck the string if you wanted to produce an especially loud sound? Were you careful not to let this way of changing the loudness influence your comparison of the sounds from strings vibrating in two different directions?

When you installed the sixth string, you probably noticed that its sound changed as the string was tightened. Let us now find out how the sounds of guitar strings are affected by tightening the strings or shortening them.

Remove the pencil and hold the guitar so that you can pluck the guitar strings. Pluck the six guitar strings, one at a time. A freely vibrating guitar string is referred to as an *open* string.

Pitch is a word we use to differentiate musical tones. We say that one musical tone is higher or lower in pitch than another. Or, the two tones may have the same pitch. For example, a woman's voice usually has a higher pitch than a man's. The pitch produced by plucking guitar strings becomes higher from largest string to the smallest string.

Pluck any one of the six strings and listen to its pitch. Shorten the portion of a string by pressing it with your finger so that the string is held firmly against the first fret below the nut. Now pluck the string again. Continue to shorten the string one fret at a time, pluck the string, and listen to the pitch.

8. Does the pitch go up or down as the string is shortened? Does the same result hold for all strings?

Pluck one of the strings with a guitar pick at a point over or near the sound hole. We call this sound *rich* or *full*. Next pluck the same string near the bridge. The sound produced is called *tinny*. Finally, pluck the same string exactly at its middle. (The midpoint is halfway between where the string crosses the nut and where it crosses the bridge.) This sound is called *hollow*. Pluck the string at each of these positions in succession. Can you distinguish between these different sounds? Rich, tinny, and hollow are examples of words used to describe the *quality* of sound.

Have your lab partner pluck another string to produce one of these three kinds of sounds. Turn your back. Ask him to pluck the string in different places until you can correctly identify where he is plucking it each time.

Pluck a string near the sound hole with a guitar pick and with your finger or thumb.

9. How is the quality (rich, tinny, hollow) of the sound affected by using a pick?

Strum the fifth string with the sound hole open, and again when it is covered with an index card. You may want to cover and uncover the sound hole a few times while one of the strums is sounding.

10. Describe any difference you hear in tone quality.

With the hole covered, pluck the string at the place that would ordinarily result in a rich sound. Then uncover the hole and pluck the string so as to produce first a tinny sound and then a hollow sound.

11. Does covering the hole cause the sound produced by a normal strum to change from rich to either tinny or hollow?

Pluck each of the other guitar strings near the sound hole, while covering and uncovering the hole.

12. Does the open sound hole emphasize any particular open string sounds more than others? If so, which strings are most strongly emphasized?

Pluck a string near the sound hole so as to produce transverse vibrations that are parallel to the surface of the guitar fingerboard. Then pluck the string to produce vibrations that are perpendicular to the fingerboard.

13. Is there any difference in the quality of these two sounds?

Let us now see how the tension (tightness) and diameter of a guitar string affect its pitch. Pluck the string which has the lowest pitch. By turning the tuning peg to which it is attached, increase the tension on the string and pluck it again.

14. What effect does increasing the tension of a string have on its pitch?

Following the instructions provided with your guitar, tune each of the strings.

At the midpoint of the strings, tape a ruler under the strings for use as a scale, as shown in Figure 9. Cut a small rubber band so that you can loop a single strand around a string. Slip the rubber strand around the first string above the ruler and grasp both loose ends between your thumb and forefinger. You can now displace the string from its normal position by pulling along the ruler on the rubber strand, and you can measure the displacement by sighting down on the ruler.

Use the rubber band to displace the midpoint of each string a fixed amount (say ¼ in). Measure the length of the stretched rubber strand when it is holding a string in

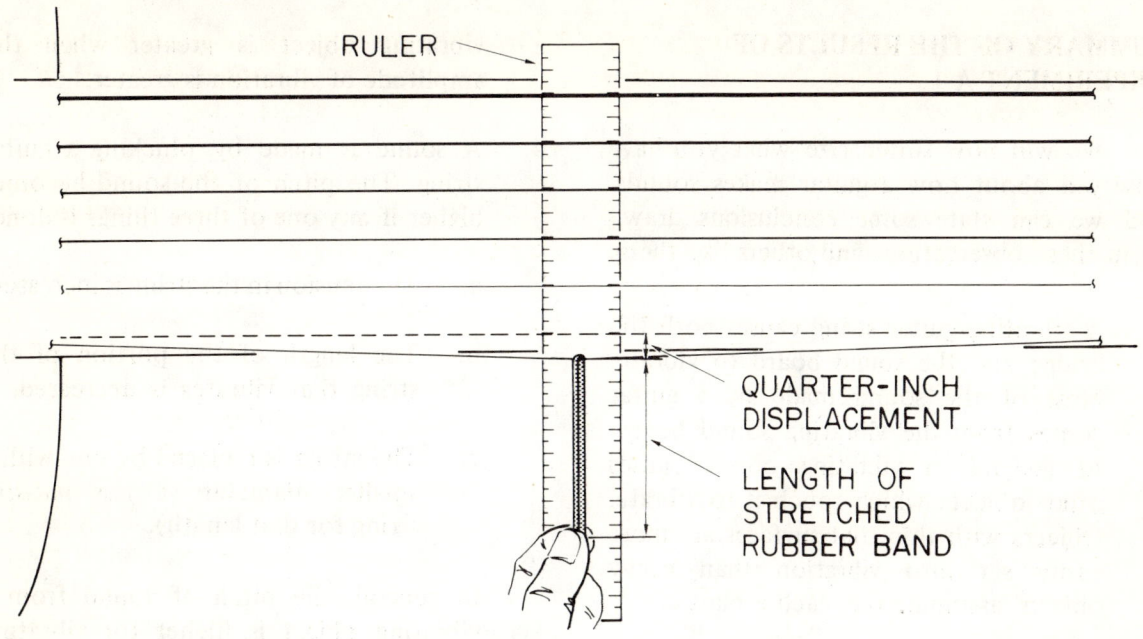

Figure 9.

this displaced position. You can do this by taking the reading on the ruler where you are holding the open ends of the rubber band and subtracting it from the reading on the ruler at the place where the rubber band pulls on the guitar string. This length is a measure of the force required to hold the string in its displaced position, and this force increases with the tension in the string. Thus by comparing the lengths of the stretched rubber strands required to hold the various tuned strings in equally displaced positions, you can compare the tensions in the six tuned guitar strings.

15. How do the tensions compare? Are they about the same or quite different?

Since the notes produced by plucking guitar strings of equal length have different pitches even when the string tensions are approximately equal, something about the strings themselves must affect the pitch.

16. What difference is there between any two of the six strings? How does this factor seem to affect pitch?

SUMMARY OF THE RESULTS OF EXPERIMENT A-1

We will now summarize what you have observed about how a guitar makes sounds, and we can state some conclusions drawn from these observations and others like them.

1. A vibrating guitar string causes both the bridge and the sound board to vibrate. Most of the sound made by a guitar comes from the vibrating sound board. In general, a vibrating object causes other objects which touch it to vibrate. Objects with thin, flat surfaces are more easily set into vibration than heavy objects, assuming that each is elastic.

2. You observed that a wire connecting one tin can to another transmits sound from one vibrating can bottom to the other by vibrating longitudinally. You also saw that the sound board of a guitar produces sound by vibrating in a direction perpendicular to the surface of the board. You found that the sound board vibrations are most readily excited when the strings vibrate transversely in a direction perpendicular to the surface of the sound board. You also observed that longitudinal and transverse pulses move along a coil at *different* speeds. In general, springs and wires can be made to vibrate in either a transverse or a longitudinal manner, or a combination. The speeds at which transverse and longitudinal pulses move are not the same.

3. The sound made by a guitar is loud when the string that excites the sound is displaced a large distance from its rest position before it is released. The maximum displacement at a point on a vibrating string is called the *amplitude* of the vibration at that point. In general, the loudness of a sound produced by a vibrating object is greater when the amplitude of vibration is greater.

4. A sound is made by plucking a guitar string. The pitch of the sound becomes higher if any one of three things is done:

   a. The tension in the string is increased.

   b. The length of the portion of the string that vibrates is decreased.

   c. The string is replaced by one with a smaller diameter (a less *massive* string for that length).

   In general, the pitch of sound from a vibrating object is higher for vibrating objects of small dimensions, it is higher when forces required to displace the object are large, and it is higher for vibrating objects of small mass, but having the same dimensions.

5. The quality of sound made by a guitar is influenced by these factors:

   a. The place where the strings are plucked or strummed

   b. The object used to pluck the strings

   c. Whether the sound hole is open or closed

   In general, the same factors apply to any stringed instrument. Other factors, more difficult to investigate, such as materials and construction, also affect the quality of sound from stringed instruments.

6. The open sound hole helps to increase the loudness of the sound made by a guitar and to make it richer. This effect is more noticeable for low-pitched sounds than for high-pitched sounds.

VIBRATIONS

What causes an object to vibrate?

A tuned guitar string that has not been plucked is at rest. The string is under *tension*. Every piece of the string is being pulled by adjacent parts of the string. A particular piece does not move because the pulls exerted by its neighbors are equal in magnitude and opposite in direction (see Figure 10). The result is that any piece of the string feels no net force.

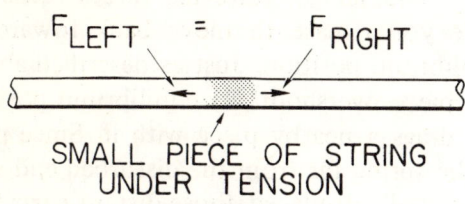

Figure 10.

If you grasp a piece of string and pull it away from its rest position, neighboring pieces of string are pulled aside also, but not as far. As shown in Figure 11, the two neighboring pieces of string now try to pull the piece you are holding toward its original rest position. The two forces are called *restoring forces* because they tend to restore the string to its original position.

When you let go, the restoring forces shown in Figure 11 cause the string to move back toward its rest position. The forces exerted by each side no longer cancel each other. Instead they act together to pull the string back toward its rest position.

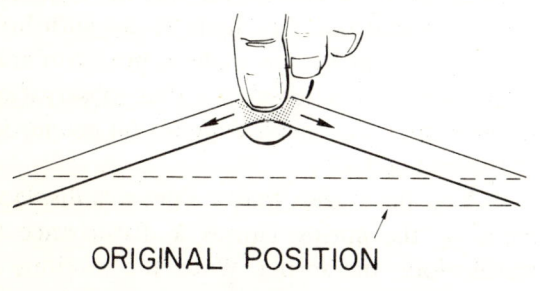

Figure 11.

The harder a string is pulled initially, the greater is the displacement of the string. The greater the initial displacement, the greater are the restoring forces; therefore, when it is released, the string will be moving faster when it reaches its original rest position (called the *equilibrium position*).

Because the string is moving when it reaches the equilibrium position, it moves on past. As soon as the piece of string you plucked has moved past the equilibrium position, its adjacent parts begin to pull it back toward the equilibrium position as shown in Figure 12. The restoring forces, reversed in direction compared to those shown in Figure 11, slow the string down and bring it to rest with a displacement opposite to its initial displacement. Then it moves again toward the equilibrium position. These processes repeat over and over, resulting in the back-and-forth motion that we call vibration.

Figure 12.

Question 1. Suppose that you suspend a weight on the end of a vertically held spring. When the weight is displaced downward and released, it will vibrate up and down. Describe this vibration in terms of restoring forces.

Question 2. Describe how restoring forces produce vibrations in some other physical situations: the bottom of the tin can in the "telephone" we discussed earlier, a ruler clamped to the edge of a table, and a tine of a tuning fork.

How does a vibrating string cause other things near it to vibrate? If the string were tied down at its ends to *immovable* support points, then only the string would vibrate and little sound would result. Suppose instead that one end of the string is attached to a structure that can move; for example, a bridge and sound board of a guitar. Then whenever forces caused by tensions in displaced pieces of string pull up or down on the bridge, it moves a little in response to these forces, just as pieces of the string do. The resulting motion is an up-and-down vibration of the bridge. The vibrating string *excites* the bridge and sound board.

The ease with which the bridge and sound board can be excited depends on two factors which are common to all vibrating structures.

The first factor which helps to determine how a structure vibrates is its *elasticity*. An object is elastic if when it is deformed (its shape is changed) there are forces that tend to restore the object to its original condition. A putty ball is not elastic and does not vibrate well. On the other hand, steel is highly elastic, and a steel rod readily vibrates, in spite of the fact that the steel rod is harder to deform. In fact, steel, bronze, and nylon, the common materials from which guitar strings are made, are much more elastic than a more familiar "elastic" material such as rubber.

Question 3. Which is more elastic, a strip of lead or a strip of steel with the same dimensions?

Small forces can cause displacements of the sound board of a guitar from equilibrium. The sound board is elastic enough that when excited, it will vibrate quite a few times before coming to rest.

A second important factor is the *mass* of the vibrating structure. It is common experience that massive objects are hard to move. If a large mass were attached to the sound board of a guitar, it would be harder to make the sound board move back and forth. The vibration that would be easiest to excite would then be different—less rapid and perhaps involving less overall motion. Thus the associated sound would be different, both in pitch and loudness.

TRANSVERSE AND LONGITUDINAL PULSES

So far we have talked only about the vibrations of an object as a whole. What happens if a disturbance is created at one location in a large elastic object which is capable of vibrating? You learned some answers to this question when you pulled aside a piece of the long spring near one end and then released the part you were holding. A neighboring piece of spring farther from the end must follow the piece you move first. When you let go, restoring forces cause the piece you release to move back toward its equilibrium position. Just as described above, this piece overshoots its equilibrium position and drags a nearby piece with it. Since parts of the spring far from the disturbed end were not initially displaced, those distant parts have not yet discovered that anything is going on at your end of the spring. However, as soon as a piece of spring a short distance away from the end is pulled past its equilibrium position, it pulls a piece a little farther away from the end after it, and this second piece in turn drags a third piece somewhat farther from the end after it, and so on. As time passes the disturbance travels farther from the end where it first started. The bump or shape that moves along the spring is called a "wave pulse." Note that what moves from one end to the other is this disturbance and not chunks of material, as indicated in Figure 13.

What determines how fast this pulse moves along the spring? That is, what determines the *speed* of the pulse? You probably did not make enough observations to answer this question. However, you may have noticed that the speed has very little to do with how far you pull the spring aside or how you hold it before you release it. This observation suggests that the speed depends on properties of the spring itself.

You have seen that a sideways displacement of the spring causes a disturbance to travel along the spring. When the motions of the pieces of spring are perpendicular to the

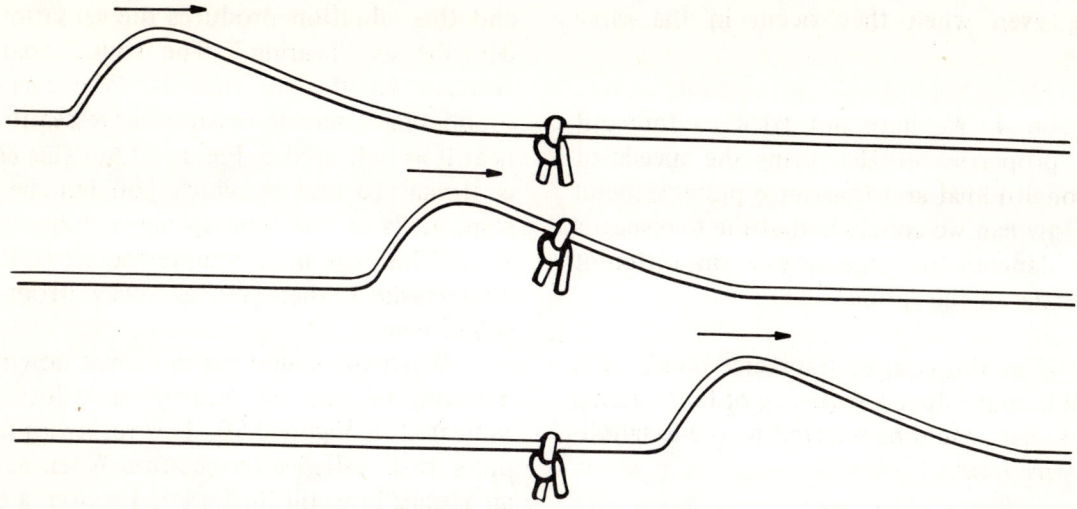

Figure 13. The wave pulse moves along the spring, but the cloth marker only bounces up and down as the pulse passes.

direction of motion of the wave disturbance (as in this case), the moving disturbance is called a transverse pulse. If several such disturbances follow each other, and successive pulses are separated by equal time intervals and alternated in direction so that one is to the left and the next one to the right as shown in Figure 14, we have what is called a *transverse wave train,* or simply a *transverse wave.*

When several coils near one end of the spring are bunched up, then released, another kind of disturbance moves along the length of the spring. In this case the forces that tend to cause a displaced piece of spring to move back toward its equilibrium position are directed parallel to the length of the spring. Again, it happens that the displaced piece passes beyond its equilibrium position. This produces a bunching up of coils a short distance away from the end. This compressed part of the spring pushes on coils of the spring that are still farther from the end. The result is that the disturbance moves down the spring away from the end where it started. Since the vibration of matter (coils) in this case is along the same direction as the motion of the disturbance, we call this a *longitudinal pulse.*

The results you obtained when you produced simultaneous transverse and longitudinal pulses in the spring showed that the speeds of these two pulses are not always the

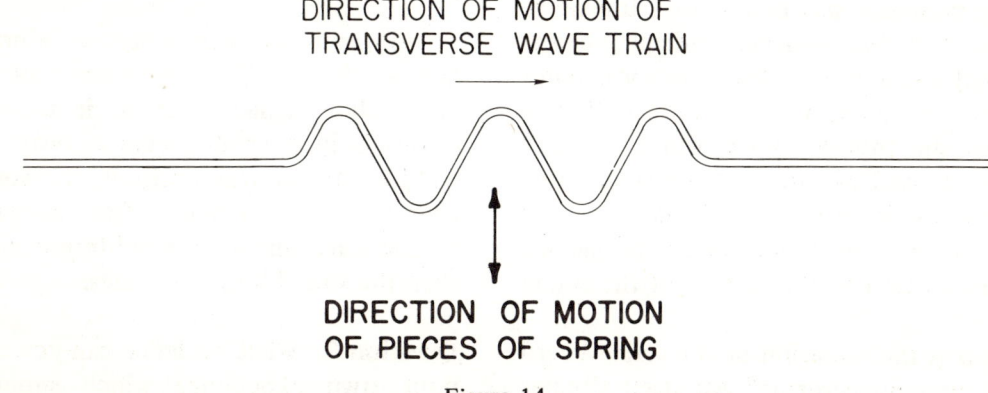

Figure 14.

same, even when they occur in the same spring.

Question 4. We have not tried to find out what properties of the spring the speeds of the longitudinal and transverse pulses depend on. How can we conclude that the two speeds *must* depend to some degree on different properties of the spring?

As in the case of transverse pulses, it is possible to produce a sequence of longitudinal pulses that would be referred to as a *longitudinal wave train.*

Question 5. Roughly sketch the appearance of a longitudinal wave train, as it would appear moving along a spring like the one you used in Experiment A-1.

VIBRATIONAL COUPLING

When one vibrating structure excites another, we say there is *vibrational coupling* between them. If the material of the second structure vibrates weakly, even though the first medium vibrates strongly, we say that the coupling is *weak*. Whether the coupling is weak or strong depends on the direction of vibration of the original wave.

For example, the sound board of a guitar is sufficiently elastic only in the direction perpendicular to its surface. We cannot excite vibration of the sound board by coupling the bridge to a guitar string that is vibrating only in the longitudinal direction. You also noted that the coupling was better for a transverse vibration in a direction perpendicular to the surface of the sound board than for a transverse vibration parallel to that surface, though the latter did produce some sound. Much of this sound resulted from unavoidable vibrations of the string in a perpendicular direction rather than from vibrations of the bridge in a direction parallel to the surface of the sound board.

How is the vibration of the sound board coupled to your eardrum? Air itself vibrates and this vibration produces the sensation we describe as "hearing." The sound board is coupled to the air near it. Whenever the sound board moves up, it compresses the air near it as indicated in Figure 15A. (This effect is similar to that in which you bunched up some coils of the long spring in Experiment A-1.) This region of compressed air (called a *compression*) then moves away from the sound board.

When the sound board moves down, the pressure of the air nearby is reduced, as indicated in Figure 15B. This region of lower pressure is called a *rarefaction*. When nearby air rushes in to fill this rarefied region, a pulse of rarefaction moves away from the sound board. As shown in Figure 15C, a succession of alternate compressions and rarefactions constitutes a longitudinal wave train, called a *sound wave,* that moves away from the sound board in all directions. When these longitudinal waves reach an eardrum, the alternating pulses of higher and lower air pressure cause the eardrum to vibrate just as the bottom of the tin can was caused to vibrate by the wave pulses in the wire. The eardrum is thus coupled to the vibrations that occurred at the sound board at a somewhat earlier time.

Question 6. Using a sketch, describe how compressions and rarefactions of air moving longitudinally from your mouth to the inside of a tin-can telephone make the tin can bottom vibrate.

LOUDNESS AND AMPLITUDE

If your eardrum does not vibrate, you hear no sound. If the amplitude of eardrum vibration is small, that is, if the eardrum moves only a small distance away from its equilibrium, or rest position, the sound you hear is "soft," or quiet. If the displacements of the eardrum from equilibrium are larger, then the sound heard is louder.

Question 7. What evidence can you cite from your own experience which supports the

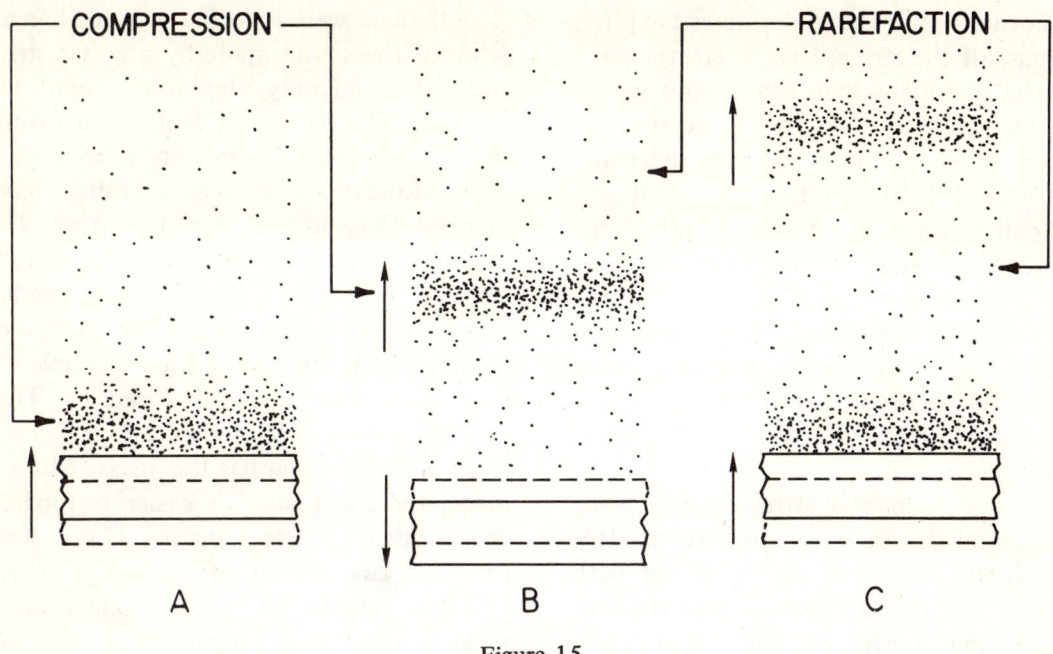

Figure 15.

statement that loud sounds result from large initial displacements from equilibrium?

If you cause a large displacement of a guitar string by plucking strongly, then that point and all neighboring points vibrate (or *oscillate*) with larger amplitudes than would have resulted if you had plucked gently. As a result of a large initial displacement, the vibrations of the bridge and sound board are more vigorous, and the amplitude of the resulting sound wave that reaches your ear is larger. This, in turn, causes your eardrum to experience large displacements from equilibrium and the sound you hear is loud. The larger the amplitude of vibration of a sound source, the greater the loudness heard at some fixed distance from the source.

FREQUENCY AND PITCH

To play a musical instrument, a musician must be able to control, at will, the pitch of the sounds that his instrument makes. For stringed instruments, there are three ways to control the pitch. The guitarist uses all three of them.

First, you already know that increasing the tension in a vibrating string raises the pitch of the sound that it makes. One end of each guitar string is wrapped around a peg. The peg can be rotated by turning a gear system with your fingers. In this way the tension, and hence the pitch, of each string can be adjusted as desired. Making these adjustments is referred to as *tuning the guitar*.

Why does increasing the tension in a guitar string raise the pitch? The tension in the string supplies the restoring forces that cause the string to vibrate. The stronger the force, the greater the rate at which the string vibrates. If the string vibrates more rapidly, the number of *oscillations* (vibrations) occurring in each second increases. We call the number of complete back-and-forth oscillations per second the *frequency of oscillation* (*vibration frequency*). The frequency increases as the tension increases. If it can be established that the pitch also increases with frequency, the observation that pitch increases with tension will be easier to understand. You will study the relationship between pitch and frequency in Experiment A-2.

Another factor that determines the pitch is the mass of the string. Massive strings make lower-pitched sounds than less massive strings of the same length under the same tension. This fact is used to help the guitarist make sounds with different pitches. The six strings on a guitar have the same lengths, but different thicknesses, and therefore different masses. When the guitar is tuned, all the strings are under approximately the same tension. The six strings, when plucked, make sounds that have six different pitches. These differences result from the different masses of the strings.

Why does a massive string make a lower-pitched sound than a less massive string? It is more difficult to speed up an object with large mass than one with small mass. It seems reasonable that a given restoring force causes a massive string to vibrate less rapidly (with lower frequency) than a less massive string. Again, you might suspect that pitch is related to frequency of oscillation.

If there were no other way to change the pitch of the sound made by a guitar string, a guitarist could only play songs composed of six particular notes. In fact, on instruments like a piano, each string produces only one note. However, there is a third way to increase the pitch of a guitar string. This is done by shortening the part of the string that vibrates. Why does a shorter string produce a higher pitch? You might be tempted to answer that this is just another example of the mass effect that we discussed earlier. That is, you might say that since a short piece of a given kind of string has less mass than a long piece, the short piece is easier to move. But this is not the correct answer. If you were to compare two strings of the same mass but different lengths, the shorter one would still make a sound with a higher pitch. Why does a shorter string produce a higher pitch? The following experiment will help you answer this question.

EXPERIMENT A-2. Harmonics and Modes of Oscillation

String musicians use a particular aspect of string vibrations called *harmonics*. In this experiment you will learn how to produce harmonics in a vibrating string. You will also observe these same harmonics in a long spring.

Place a strip of masking tape along the fingerboard beneath the first string of the guitar. On this tape, number the frets starting at the nut end of the neck. Place another strip of tape across all but the sixth (largest diameter) string of the guitar. This will keep the other strings from vibrating. The arrangement is shown in Figure 16.

Pluck the guitar string at its midpoint with your thumb.

1. Study the appearance of the vibrating string; then sketch how the amplitude of vibration varies from one end of the string to the other. This dominant pattern of vibration is called the *fundamental* or *first harmonic*. To learn more about the first harmonic, we will look at a larger vibrating system which *simulates* the guitar string.

Horizontally suspend the spring provided for you as shown in Figure 17 so that there is a distance of three or four meters between the ends of the spring. The ends should be clamped so that they cannot move, and the sag at the center should be less than 30 cm. Measure the length of the spring and, with small pieces of tape, mark points that are at distances of one-half, one-third, one-fourth, and one-fifth of the length of the stretched spring from one end.

Pull the spring aside at its midpoint. Then release it.

2. How does the vibration pattern of the spring compare with that of the guitar string?

Practice swinging the spring in time with its natural vibration. To do this push it to one side then the other at a point near one end of the spring, always keeping time with its natural vibration rate. You should be able to produce large vibrational motions (large *amplitudes*) by always pushing when the spring is moving away from you and pulling back when the spring changes direction and starts to move back toward you again. The timing is similar to pushing a child's swing to make the child swing higher. Supplying such a *periodic* force to a vibrating system is called *driving* the system.

3. Does the amplitude of the vibration seem to affect the timing of the vibrations (the frequency)?

Now go back to the guitar and pluck the sixth string hard, then more softly.

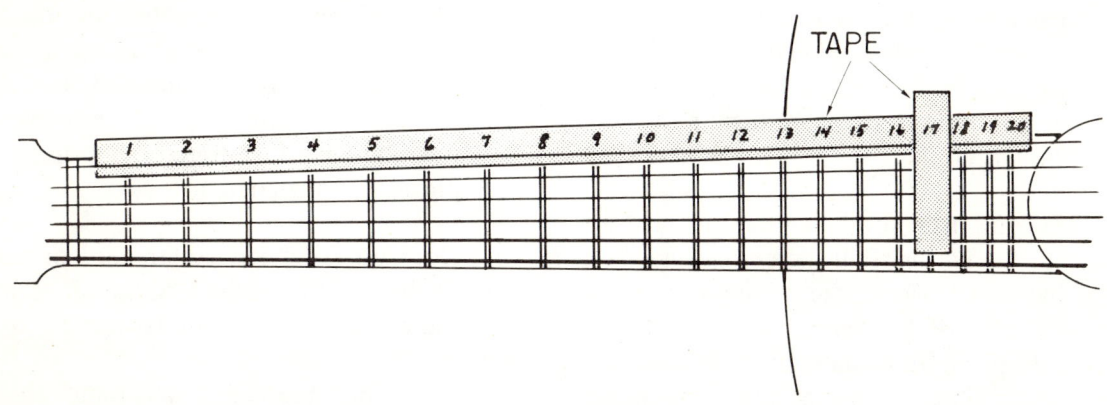

Figure 16.

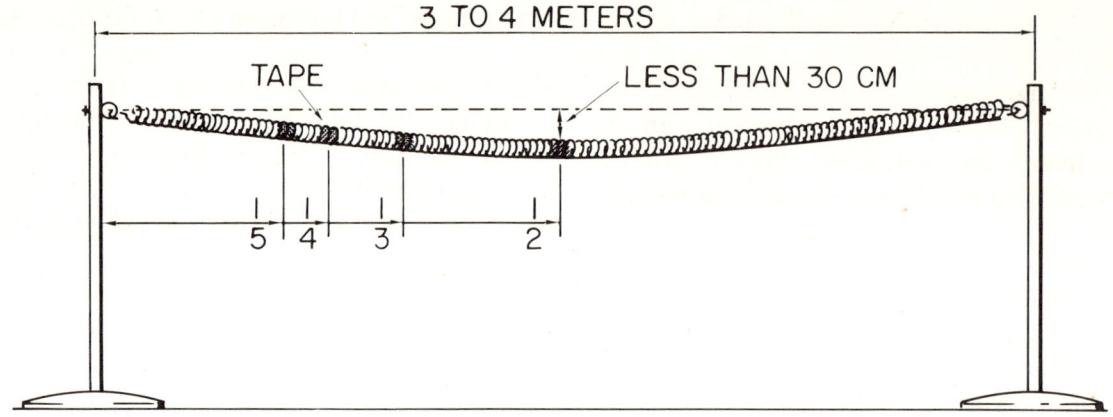

Figure 17.

4. Does the pitch change when the string is plucked more softly?

To produce a different but related vibration, called the *second harmonic,* pluck the guitar string at a point over the sound hole. We will call this the *normal* position for plucking a guitar string. Then *lightly* touch the midpoint of the string with your finger. The midpoint is usually located at the 12th fret. Measure to make sure you are touching the midpoint.

5. How does the pitch you hear after touching the string compare with that you heard before? (If you hear nothing after touching the string, you probably are touching the string with too much force.)

6. You probably can't see the variation in string amplitude for the second harmonic, but try to determine how the string is vibrating by *lightly* touching the string at various points after producing the second harmonic.

We can also produce the second harmonic in the spring. Pull the spring aside at a point halfway between the midpoint and one end, and release it. Then carefully catch and stop the spring at its midpoint. Release the spring immediately and observe the remaining vibrations.

7. Sketch the variation of the amplitude of the spring vibration along its length.

By pushing back and forth on the spring at a point near one end in time with the second harmonic, you can increase the amplitude of this type of vibration. You must take care that you introduce no amplitude at the midpoint of the spring. Such a point, where the amplitude of vibration is zero (where the spring does not move), is called a *node*.

8. How does this vibration compare with the second harmonic of the vibrating guitar string?

9. Does changing the vibration amplitude seem to change its frequency?

10. How does the frequency of the second harmonic compare with that of the first harmonic? (You can determine the frequency by counting the number of times you push the spring back and forth in a certain time interval. The frequency is the number of complete back-and-forth motions per second.)

11. Does plucking the guitar hard or softly to sound the second harmonic change the pitch of the second harmonic?

To sound the third harmonic on the guitar, pluck the sixth string at the normal position, then lightly touch the string at a point one-third the length of the string from the nut. This point is at the seventh fret.

12. How does this pitch compare with that of the first and second harmonics?

13. Can you produce this same pitch by touching a point one-third the length from the bridge end of the string?

As before, use the spring to more clearly observe the third harmonic. Pull the spring aside and release it. Stop its motion at a point one-third its length from either end. Release the spring and observe the resulting vibrations.

14. Sketch how the amplitude of the vibration varies along the length of the spring.

Drive the spring near its end as you have done for the other two harmonics to produce a large-amplitude third harmonic. Now observe the *two* nodes between the ends of the spring. Where are they?

15. How does the frequency of this third harmonic compare with those of the first and second?

To produce the *fourth* harmonic, where will you have to touch it after plucking? Try it and see if it works. (You may have to pluck the string at a point nearer the bridge to hear the fourth harmonic.)

16. How does the pitch of the fourth harmonic compare with those of the other harmonics?

17. The fourth harmonic can be produced by touching the string at a different point. Where is this point? Try it to see if it works.

While the fourth harmonic is sounding, touch the midpoint of the string.

18. Can you still hear the fourth harmonic? Try the same thing with the third harmonic.

To produce the fourth harmonic in the spring, pull it aside near one end and stop it at the correct point. (Where is this point?) Release the spring and observe its vibrations.

19. Sketch how the amplitude of the vibration varies along the length of the spring.

20. How do the frequencies of the four harmonics compare?

Produce the fifth harmonic in the guitar string.

21. How does the resulting pitch compare with the pitch of the other harmonics?

22. Can you produce the fifth harmonic by touching points that are two-fifths, three-fifths, or four-fifths the length of the string from the nut?

Produce the fifth harmonic in the spring.

23. Sketch the resulting vibration.

Drive this fifth harmonic to large amplitude as you have done before.

24. How does the frequency of this vibration compare with that of the previous harmonic?

25. How would you produce the sixth, seventh, or any other harmonic in the guitar string?

26. Do you think the position at which the string is plucked affects the number of harmonics generated and their amplitudes?

Try producing the second harmonic after plucking the string at various points.

27. Did you pluck at any points and find that you could *not* produce the second harmonic? If so, where were these points?

Try producing the third harmonic after plucking at various points.

28. Did you pluck at any points and find that you could *not* produce the third harmonic? If so, where were these points?

SUMMARY OF THE RESULTS OF EXPERIMENT A-2

You have observed the appearance of a guitar string sounding its lowest-pitched sound. A multi-exposure photograph of the string made at very short time intervals might look like Figure 18, with each line being a different position of the string. We call the resulting tone the *fundamental* or *first harmonic*. This is also the pattern of transverse vibrations in a spring when it is vibrating at its lowest possible frequency.

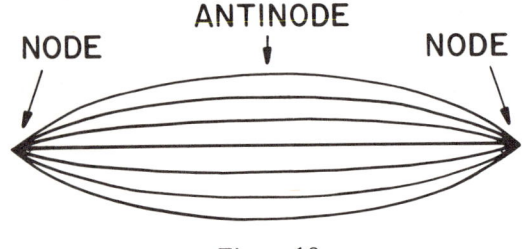

Figure 18.

When a guitar string vibrates so as to produce the pattern shown in Figure 19, the pitch of sound emitted raises to an *octave* above the fundamental. We call this higher pitch the *second harmonic*. When a spring vibrates in this pattern, its frequency of vibration is two times its lowest (fundamental) frequency. Thus, a one-octave increase occurs when the frequency of vibration is doubled.

Figure 19.

We call a point where no vibration occurs a *node*. The points of maximum vibration amplitude are called *antinodes*.

Question 8. Where are nodes located in the string vibration shown in Figure 19? Where are the antinodes located? How many of each are there?

Question 9. If Figures 18 and 19 refer to the same guitar string, how does the distance between nodes in Figure 18 compare with the distance between nodes in Figure 19? How do the frequencies compare?

Whenever the pattern of vibration on a guitar string changes in such a way as to add another antinode, and thus another node, the pitch of the sound becomes higher. When a spring vibrates in a pattern that has three antinodes, the frequency of vibration is three times as great as the lowest possible frequency. We would conclude that a guitar string does the same thing.

Question 10. A string vibrates in a pattern which has seven antinodes. Sketch a multi-exposure "photograph" of this pattern. How much higher is the frequency of this oscillation than when the string vibrates at its fundamental? How many nodes are there?

The pitch of a sound is determined by the frequency of the vibration that produces the sound. Also, there is a relationship between the frequency of vibration of a string and the distance between nodes. When the distance between nodes is halved, the frequency is doubled.

When a guitar string is plucked at the normal position, or at almost any randomly chosen point, many harmonics are produced. However, a harmonic that corresponds to a pattern of vibration with a node at the excitation point cannot be produced. For example, you cannot produce the second harmonic by plucking the string at its midpoint.

The *quality* of sound from a guitar varies with the point at which the string is plucked. This suggests that the quality of a sound is related to the harmonic content of the sound. That is, whether you hear a rich sound or a tinny sound depends on the number of harmonics present and the amplitude of each. In music theory, the word "harmonic" refers to a tone made up of all of the complex vibrations which are left after a string is plucked and then touched at some point. In physics, "harmonic" refers to a single simple vibration. You will study this aspect of harmonics more carefully later in the module.

NODE-TO-NODE DISTANCES AND FREQUENCIES OF OSCILLATION

Experiment A-2 provided evidence that a different frequency is associated with each simple pattern of vibration. You saw the similarity between the pattern on a guitar string that had been plucked at its midpoint and a long spring vibrating so that only its ends are nodes. When the guitar string is plucked at some other point, such as the normal position, almost the same basic pattern of vibration is visible. However, if the string is plucked at its normal position and then touched at the midpoint, vibration continues but the pattern changes to that associated with the second harmonic (Figure 19). The pitch increases abruptly when this change takes place. The new pitch is described by musicians as being one octave higher.

When the long spring is vibrating and it is then held at its midpoint so as to force the midpoint to become a node, the two halves of the spring may continue to vibrate in a pattern that looks much like that shown in Figure 19. The frequency of this new pattern is double the previous one. The conclusion is that spring (and string) patterns with more antinodes vibrate at higher frequencies and the strings produce sounds of higher pitches.

You also noted that for each higher harmonic, one more node is added and, for a given harmonic, the nodes are always equal distances apart. Why are these patterns possible and why do we not observe more irregular patterns? Also, why does the frequency increase when the node-to-node distance decreases? These are questions we will seek to answer in later sections of this module.

Finally, you observed that the particular mixture of harmonics excited by plucking a string depends on the details of *how* the string is plucked. Can we predict the mixture that will result from any particular position of the string just before it is released? We cannot give a complete answer to this question now, but we will come back to it later in the module. However, some aspects of the situation are clear. If you want to produce the second harmonic, or the fourth harmonic, or any even-numbered harmonic, you must *not* cause the midpoint to move, because the midpoint is a node for even harmonics. If you want to produce the third harmonic, you must be sure that the points one-third of the length away from each end are at rest, etc.

THE QUALITY OF SOUND AND PATTERNS OF VIBRATION

Associated with each pattern of vibration of a string—or any vibrating structure—is a particular pitch. Depending on how the vibration is started, the overall pattern may be complex in the sense that it includes a mixture of simple patterns (harmonics). When this occurs, the *pitch* that the human ear recognizes is the pitch associated with the lowest harmonic that is present to a significant degree. However, the *quality* of the sound is determined by the particular harmonics present, depending both on the number of harmonics that are included and on the amplitudes of these harmonics.

Why does strumming the guitar at different places produce sounds of different quality? It is because the particular mixture of patterns present in a vibrating string depends on how the vibration is produced. Also, the difference in harmonic content accounts for the different sounds produced by different instruments when playing the same tones.

Question 11. What harmonics are *not* produced when a string is strummed at a point one-quarter of the way from one fixed end?

If the presence and absence of harmonics determine the quality of a sound, why does a note stimulated by a pick sound different from the same note stimulated by plucking with your fingers? The shape of the string when it is first released after being displaced determines the amplitudes of the various harmonics that are present. The shape of a string that is pushed aside by a rounded finger is different than the shape of a string that is pushed aside by a point pick.

You have seen that the qualities of the sounds made by a guitar when the sound hole is closed and when it is open are different, at least for low-pitched sounds. What does the sound hole have to do with the vibration patterns of the guitar? A vibrating structure causes other structures with which it is coupled to vibrate. The air inside the body of the guitar is caused to vibrate by the sound board and, in turn, helps the sound board to vibrate. However, the simple vibration patterns for one structure (like a guitar string) may be quite different from those for a nearby structure (like a sound board) forced to vibrate at the same frequency. When this occurs, vibrations in the first structure do not excite large-amplitude vibrations in the second structure.

It is the job of a guitar designer to make sure that all the principal structures—the strings, the sound board, and the air in the sound box—vibrate at the same frequencies. While this goal is partially achieved in every guitar, perfect matching at all frequencies is not possible. For example, the air inside the body of the guitar vibrates more easily at low frequencies than at high frequencies. Thus the quality of low-pitched sounds is influenced to a greater degree by the vibrations of the air in the sound box. When the sound hole is closed, the coupling of vibrations of air inside the sound box to the air outside is destroyed. The mixture of patterns producing the sound is altered, and the quality of the sound changes.

THE SHAPE OF THE SOUND BOARD AND THE LOCATION OF THE BRIDGE

We have already mentioned the influence of the size of a vibrating structure on the pitch of the sound it produces. We also discussed the influence of the method of exciting a vibration on the mixture of simple patterns of vibration that are stimulated and hence on the quality of the sound. Since the sound board is the guitar structure that is coupled most strongly to the air, which in turn transmits sound vibrations to our eardrums, it is important to understand what the sound board does.

The sound board, like every elastic structure, is capable of vibrating. A particular sound board vibrates in various patterns that are determined by its shape and construction. As for strings, there are nodes in these patterns. However, instead of nodes at single points, as in the string, there are nodal *lines* and/or points along which there is no oscillation. For example, the outer edge of the sound board is not very free to oscillate. So it forms a nodal line.

A particular structure vibrates easily only at certain frequencies that are determined by the size and shape of the structure. You may then wonder how the sound board, which does not change its size, can amplify all the different pitches produced by the strings. (These pitches vary from string to string and change as the vibrating length of each string is changed.) Any elastic structure can be *forced* to vibrate at any frequency. This is true for frequencies different from those at which it vibrates when abruptly excited (for example, by being plucked) and left to vibrate naturally. However, the amplitudes of forced vibrations are very small at frequencies that are greatly different from the "natural" frequencies of the structure. The sound board must be stimulated at vibration frequencies that are not greatly different from those at which it would vibrate naturally.

The location of the bridge is chosen so that the vibrations of the strings are most easily transmitted to the sound board.

SUMMARY

In this section you have learned that sound is produced by the vibration of material objects. In a guitar, the ultimate source of the sound is the vibrating string. The string transmits its transverse vibrations through the bridge to the sound board, which transfers the sound to the air.

The larger the amplitude of vibration, the louder is the sound produced. The pitch of the sound from an open guitar string depends on the length of the string, the tension in the string, and the mass per unit length of the string. The pitch may be increased by making the string shorter, by increasing the tension, or by decreasing the mass per unit length. The quality of the guitar sound (rich, hollow, tinny, etc.) depends on how the string is set into vibration.

You learned that when an object is displaced from equilibrium, restoring forces in the material cause vibrations by producing an overshoot as the object returns. Any object with restoring forces is said to be elastic or to have elasticity. Also required for vibration is flexibility, which is the ability to bend or distort. An object can be elastic without being very flexible or flexible without being very elastic, but both properties are required for ease of vibration. The sound board of a guitar is elastic and flexible, so that it is easily set into vibration by the string to produce longitudinal waves in the air which your ears detect as sound.

You have seen that a string fixed at both ends vibrates in various basic patterns. These patterns are characterized by nodes (points where the string is not moving) and antinodes (points where the string is moving the most). The simplest such pattern, with a node at each end and an antinode at the center only, is called the fundamental. Basic patterns with more nodes and antinodes (still always with a node at each end) are called harmonics, and they are numbered. The harmonic number is the same as the number of antinodes present. Several such basic patterns, or harmonics, are usually present at the same time, giving rise to a more complex vibration. This is usually the case on a guitar string. A harmonic may be removed from a complex vibration on a string by touching the string lightly at the position of an antinode of that harmonic.

The fundamental has the lowest frequency of vibration for a particular string. Each harmonic has a vibration frequency that is an integral multiple of the fundamental. The frequency of the second harmonic is twice that of the fundamental. (In musical language, the pitch is defined to be one octave higher.) The third harmonic is three times the fundamental frequency, etc.

Finally, you saw that the pitch of the sound from a guitar string is determined by the lowest frequency present (the fundamental if it hasn't been removed). The proportions of the various harmonics that are present determine the quality of the sound.

GOALS FOR SECTION B

The following goals state what you should be able to do after you have completed this section of the module. These goals should be studied carefully as you proceed through the module and as you prepare for the post-test. The example which follows each goal is a test item which fits the goal. When you can correctly respond to any item like the one given, you will know that you have met that goal. Answers appear immediately following these goals.

1. *Goal:* Know the mathematical relationship between fundamental frequency of a vibrating string and tension, length, and mass.

   Item: If you start with a tuned guitar string and then turn the tuning peg until the tension is tripled, by what factor does the fundamental frequency change?

2. *Goal:* Know the definitions of power and intensity.

   Item: A sound wave enters a square window 1.5 m on a side, with a power of 3×10^{-2} W spread evenly over the surface. What is the sound intensity at the window?

3. *Goal:* Understand the concept of intensity level.

   Item: The background noise in a room has an intensity level of 20 dB. If that noise level doubles in intensity, what is the new intensity level?

4. *Goal:* Understand the whole-population hearing ability curves (Figure 27).

   Item: If "normal" hearing is defined by the bottom curve in Figure 27, what percent of the population has impaired (worse than normal) hearing?

5. *Goal:* Understand the concept of loudness level.

   Item: What intensity level of sound is required at 200 Hz to produce a loudness level of 30 phons?

Answers to Items Accompanying Previous Goals

1. 1.73

2. 1.33×10^{-2} W/m^2

3. 23 dB

4. 99%

5. 47 dB

SECTION B

QUALITATIVE OBSERVATIONS SUGGEST QUANTITATIVE EXPERIMENTS

In Section A you made observations of a guitar and other vibrating objects. In this way you discovered many things about guitar behavior. For example, the frequencies of oscillation of the strings determine the pitches of the sounds made by a guitar. These frequencies can be increased by increasing the tension in the strings, by decreasing the mass of the strings, or by decreasing the length of the strings. While these statements contain a substantial amount of useful information, they are *qualitative* in nature. You did not discover in Section A *how much* frequency changes for measured changes in string tension, mass, and length. You can gain more insight into guitar behavior, and the operation of other vibrating systems as well, by performing *quantitative* experiments in order to discover the mathematical relationship among all the factors that influence frequency.

Then the first goal is to find an equation relating frequency (f), tension (T), mass (m), and length (L) of a string; the *string equation*. Your earlier experiments were useful because they helped you to identify the important factors, or *variables*. However, you were not trying to measure things quantitatively at that time. You did not worry about the values of the variables, or how much they changed from one observation to another. You even allowed several variables to change simultaneously. That is, you did not attempt to *control* the variables.

Now is the time to be more careful and more systematic. In this experiment you can change any one of the three quantities (tension, mass, and length) without changing the other two. We call such quantities which can be varied independently of each other *independent variables*. Since the string frequency depends on the values of the independent variables, we call it a *dependent* variable.

If we want to know how one of these independent variables affects the dependent variable, we must study its effect independently of the other two. For example, to learn how tension affects frequency, one does experiments in which the tension is varied and the corresponding changes in frequency are measured, while the string mass and the string length are not allowed to change.

This means that to determine completely the string equation, one must perform three different experiments. In each case, it is necessary to measure the values of the independent variables that don't change, several different values of the independent variable that do change, and the frequency that corresponds to each of these values. This is then a controlled, quantitative experiment. The result will be an *empirical* equation, an equation determined by experiment. Later we will try to develop a theory for string vibrations. This theory will describe the fairly complex behavior of the vibrating string in terms of basic concepts that underlie many other complex situations as well. From this theory we may be able to derive a *theoretical* equation that relates the independent and dependent variables. If our theory is successful, the empirical equation and the theoretical equation will agree.

You should now do Experiment B-1.

EXPERIMENT B-1. The String Equation

In this experiment you will determine how the pitch, or fundamental frequency, of a vibrating string varies with the length, mass, and tension of the string. You will find work sheets for this experiment at the end of the module.

Start with a tuned guitar and a calibrated *audio oscillator*. (An audio oscillator is an electronic device that produces electric wave trains at any desired frequency over the range that includes audible sound. When the output of the oscillator is connected to a speaker, you can hear a *pure* tone. That is, you hear a single frequency with no harmonics.) Pluck the midpoint of the sixth (low-E) string of the guitar and change the frequency of the oscillator tone until the pitches of both sounds are the same. You must determine when they are the same by ear. An average person can match pure tones to those produced on a guitar with an accuracy of better than 0.3%. However, it is particularly difficult to tune a tone which is rich in harmonic content to a pure tone. Thus, be sure to pluck the midpoint of the string, so as to emphasize the first harmonic.

1. Read the frequency of the audio oscillator when you think the match in pitches is best. Record this frequency in the first column of Table 1. (The unit of frequency is called the *hertz* [Hz]; one hertz equals one oscillation per second. One kilohertz [kHz] equals 1000 Hz.)

Now, the first part of this experiment is to determine the dependence of frequency on mass per unit length of the string. To do so you will hold the tension and the vibrating length constant.

Remove the end of the string from its tuning peg and attach it to the spring scale as shown in Figure 20. Hold the guitar in place by using straps and clamps as shown. Put the string under tension by pulling on the scale and pluck the string at the midpoint of the part that vibrates. Make certain the string is pulled downward over the nut of the guitar. This insures that the vibrating length of string is the same as before (from the bridge to the nut). Pull on the string until its frequency again matches the oscillator frequency.

2. The reading of the spring scale is the tension in the string. Record the tension (in newtons, N) necessary to produce this oscillation frequency at the top of Table 1.

3. Remove the string from the guitar. Measure and record its mass and total length in columns 2 and 3 of Table 1. Express the mass in kilograms (kg) and the length in meters (m).

4. Divide the string mass by its length to calculate the mass per unit length, and record this result in column 4 of Table 1.

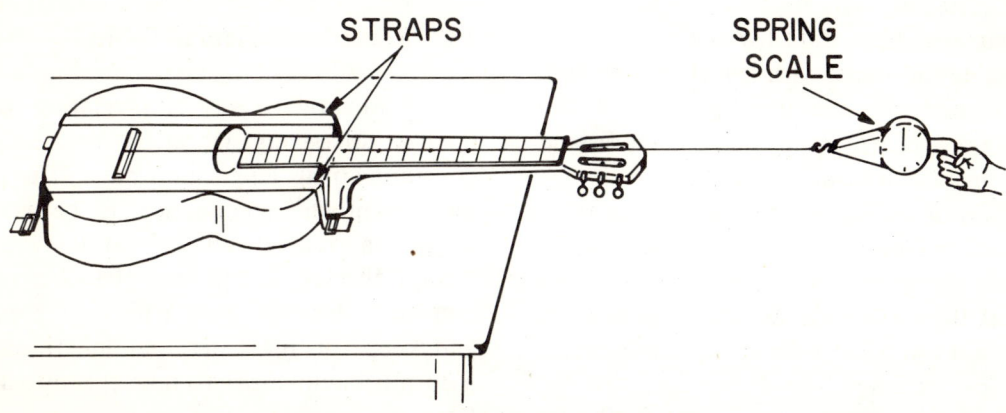

Figure 20.

5. Measure the distance from the nut to the place where the string rests on the bridge. This is the length of the part of the string which is vibrating. Record this value L at the top of Table 1.

6. Multiply the length L by the value of mass per unit length from column 4 and record the resulting mass m in column 5 of Table 1. This is the mass of that portion of the string which vibrates.

Remove the next (fifth) string from its tuning peg and attach it to the spring scale. Pull on the scale until you have the same tension as for the sixth string. Again be sure that the string is pulled tight against the nut. (You will recall that the tension is about the same for each string of a tuned guitar.) Match the oscillator frequency to that of the string.

7. Record the oscillator (string) frequency in Table 1.

8. Remove the string and measure and record its mass and total length.

9. Calculate the mass per unit length. Record this value in Table 1.

10. Calculate the mass m that vibrates using the value of L from step 5 and the mass per unit length from step 9.

Repeat this procedure for each of the remaining strings, keeping the tension the same in each case.

You now have data for the relationship of frequency to mass, with length and tension held constant. In the next section of the experiment you will keep the string mass and length constant and vary the tension.

Clamp a pulley to the edge of the table and attach a 1-kg weight hanger to the end of the sixth string, as shown in Figure 21. Be sure that the string presses firmly against the nut, so that the part of the string that vibrates is the normal length. In Table 2 record the mass of the part of the string which vibrates (from the 5th column of Table 1), and the length L of the part of the string which

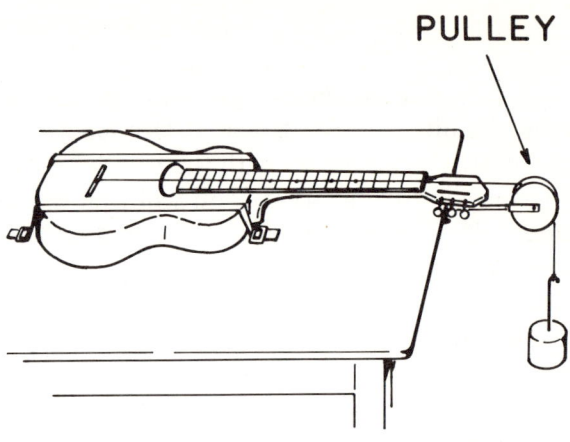

Figure 21.

vibrates. Also record the mass per unit length of this string.

Match the frequency of the string with that of the audio oscillator.

11. Record this frequency in Table 2.

In this case, the tension in the string is equal to the weight of the hanging mass. Calculate the tension (in newtons) by multiplying the mass (in kilograms) on the end of the string by the acceleration of gravity (9.8 m/s^2). That is, for a 1-kg hanging mass, the tension is 9.8 N.

12. Record this tension in Table 2.

Repeat this procedure for total masses of 2, 3, 4, 5, and 6 kg pulling on the string. (Remember to include the mass of the holder.)

13. Record your data in Table 2.

You now have data relating the frequency of string vibration to tension, at constant string length and mass. The remaining unknown dependence is between frequency and length. It is not convenient to change the length and keep the vibrating mass constant. Instead, you will vary the length and keep the mass per unit length constant. You must remember to take this into account later.

Adjust one of the guitar strings until it is approximately in tune. Record the approximate string tension and the mass per unit length at the top of Table 3. Match the frequency of the oscillator to that of the string.

14. Record the frequency in Table 3.

To change the length of the vibrating string, place the special clamp you have been provided (called a *capo*) on the neck so that the string is held tightly against the first fret. The capo should be placed close to the fret on the side nearer the nut. The new vibrating length is from the bridge to the first fret.

15. For the capo positioned near five different frets, measure and record, in Table 3, the fundamental frequency of vibration and the vibrating length of the string.

To analyze the data, we will look at the relationship of frequency to only one of the independent variables—mass, tension, or length—at a time. Because you really held mass per unit length, not mass, constant in the third part of the experiment, and because of the relationship between mass and mass per unit length, you can regard mass per unit length, tension, and length as the three independent variables.

The method you will use to analyze the data is to graph frequency versus the independent variables, looking for straight-line (*linear*) relationships. Generally, these relationships will not be linear and you will have to look for other relationships, such as the frequency versus the reciprocal or square root of a quantity. For example, suppose you had the relationship between two quantities x and y shown in the table. In Figure 22, y is graphed versus x, $\sqrt{x}$, and $1/\sqrt{x}$. Of the three, Figure 22C exhibits the only linear relationship. Thus one can write $y = a + b/\sqrt{x}$, where a and b are constants. In this case $y = 3 + 12/\sqrt{x}$.

Using this technique, there are many combinations possible so, to save you time, we shall suggest which quantities to graph.

We will consider first the relation between mass per unit length and frequency.

x	y
1	15
2	11.5
3	9.9
4	9

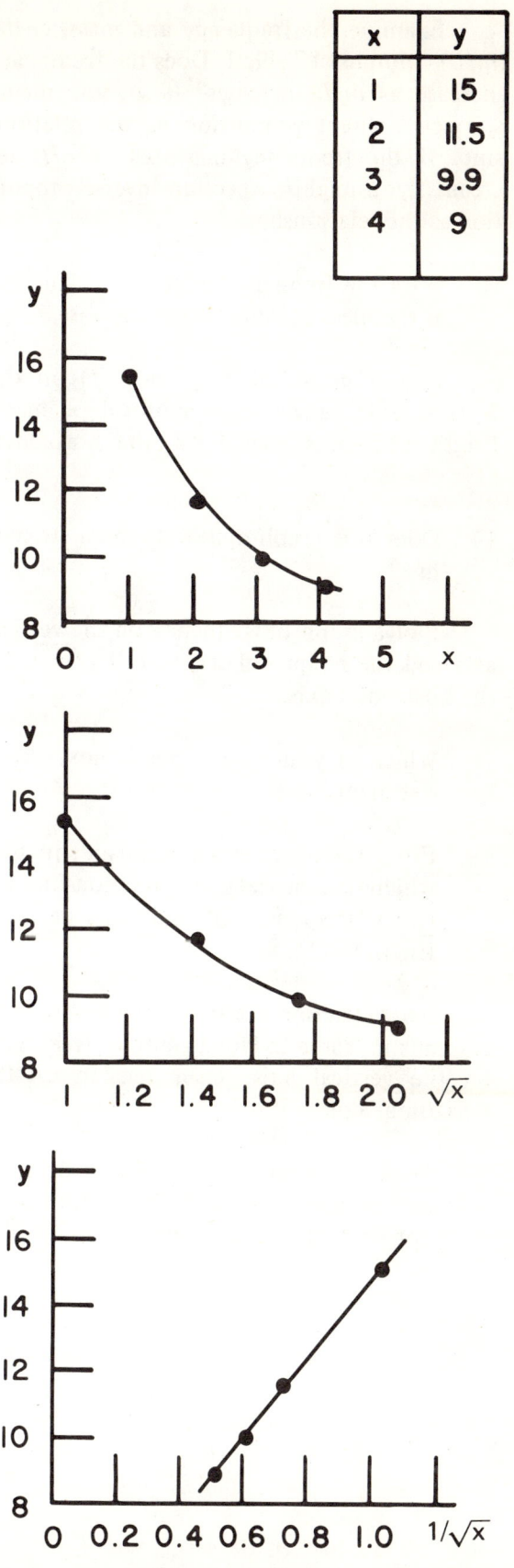

Figure 22A, B, and C.

Examine the frequency and mass/length (m/L) columns of Table 1. Does the frequency increase as m/L increases? If so, you might suspect a direct proportion as the relationship. If the frequency increases as m/L decreases, you might suspect an inverse proportion as the relationship.

16. Does the frequency increase or decrease as the mass per unit length increases?

Plot a graph of frequency (f) on the vertical axis versus the reciprocal of mass/length (1 ÷ mass/length) on the horizontal axis.

17. Does this graph appear to be a straight line?

Plot a graph of frequency on the vertical axis and the reciprocal of $\sqrt{m/L}$, $1/\sqrt{m/L}$, on the horizontal axis.

18. Which of your two graphs is most nearly a straight line?

19. From the linear graph, write an equation which summarizes the relationship between frequency (f) and mass per unit length (m/L).

Now examine the frequency and tension columns in Table 2. Plot a graph of frequency on the vertical axis versus tension on the horizontal axis.

20. Does the frequency increase or decrease as tension increases? Does this graph appear to be a straight line?

Plot a graph of frequency on the vertical axis and the square root of tension ($\sqrt{T}$) on the horizontal axis.

21. Which of these two graphs is most nearly a straight line?

22. From the linear graph, write an equation which summarizes the relationship between frequency and tension.

Now examine Table 3. Plot a graph of frequency on the vertical axis and length on the horizontal axis.

23. Does the frequency increase or decrease as the length increases?

Plot a graph of frequency on the vertical axis and the reciprocal of length on the horizontal axis.

24. Does this relationship appear linear?

25. Do you think some other relationship would be more nearly linear? If so, try it.

26. From the appropriate graph, write an equation which summarizes the relationship between frequency and length.

27. You now have relationships between frequency and mass per unit length, frequency and tension, and frequency and length. Try to express *one* relationship involving frequency, mass per unit length, tension, and length, and therefore one equation relating f, m/L, T, and L.

DISCUSSION OF EXPERIMENT B-1

You discovered in Experiment A-1 that the more massive strings on the guitar produce lower-pitched sounds, and in Experiment A-2 that low-pitched sounds are produced when the frequency of oscillation is low. You were probably not surprised when Experiment B-1 revealed that the frequency of the sound produced by the various strings on a guitar decreases as the mass of the vibrating string increases. Since it is more convenient to keep mass per unit length constant instead of mass, we regard mass per unit length as the independent variable. That presents no problem so long as the length of the vibrating string is kept constant. You again saw that frequency decreases when mass per unit length increases.

What about your results? The fact that the graph of frequency and the reciprocal of the square root of the mass per unit length is a straight line through the origin means that the frequency is proportional to the reciprocal square root of mass per unit length. Saying it another way, frequency is *inversely* proportional to the square root of the mass per unit length when the tension and string length are constant. This proportionality relationship can be written in equation form as follows:

$$f = \frac{A}{\sqrt{m/L}} \qquad (1)$$

The proportionality constant A is the *slope* of the straight-line graph of f versus $1/\sqrt{m/L}$. The value of A may depend on anything held constant during measurements of frequency and mass per unit length. Thus A might depend on the tension in the string and on the length of the vibrating string.

You discovered in Experiment A-1 that the pitch (and hence the frequency) of a guitar sound increases as the tension in the vibrating string increases. Your graph of frequency and the square root of the tension indicates that the frequency is proportional to the square root of the tension when the mass and length of the vibrating string are constant. Mathematically, this result can be expressed as follows:

$$f = B\sqrt{T} \qquad (2)$$

The constant B is equal to the slope of your graph of f versus $\sqrt{T}$. It might depend on anything held constant in this part of the experiment (the mass and/or the length).

Finally, you noted that when you shorten the length of a vibrating string its pitch, and hence its frequency of vibration, increases. If the tension and mass per unit length are constant, you found that the frequency is inversely proportional to the length. This can be expressed as follows:

$$f = \frac{C}{L} \qquad (3)$$

The constant C is the slope of your graph of f versus L. It might depend on anything held constant during this part of the experiment (tension and mass per unit length).

Now we are faced with an interesting and difficult problem. We must find a single equation that is consistent with Equations (1), (2), and (3), and thus expresses correctly all of your experimental results. You attempted to solve this problem when you answered the last item of the experiment. The mathematics needed to find such an equation in a rigorous and logical way are beyond the scope of this module. We will instead state the equation and show that it works. The correct equation is

$$f = \frac{K}{L}\sqrt{\frac{T}{(m/L)}} \qquad (4)$$

where K is a constant.

Let's compare Equation (4) with the previous equations to see if it satisfies all the requirements.

Substituting the right-hand side of Equation (1) for f in Equation (4), we get

$$\frac{A}{\sqrt{m/L}} = \frac{K}{L}\sqrt{\frac{T}{(m/L)}}$$

Multiplying both sides of this equation by $\sqrt{m/L}$, we have

$$A = \frac{K}{L}\sqrt{T} \qquad (5)$$

This means that A is constant so long as tension and length are constant. It also means that the slope of the graph of frequency and mass per unit length is larger for larger values of the constant and tension. The slope is smaller for larger constant lengths.

By similar algebraic steps, you can find that

$$B = \frac{K}{L}\sqrt{\frac{1}{(m/L)}} \qquad (6)$$

$$C = K\sqrt{\frac{T}{(m/L)}} \qquad (7)$$

Problem 1. By carrying out the necessary algebraic steps, show that Equations (6) and (7) are correct.

In each case we see that the constant of proportionality depends on the independent variables that were held constant during the appropriate part of the experiment. The constant K can be calculated from any one of the Equations (5), (6), or (7). If you calculate K from more than one of these equations, you can check the consistency of various parts of the experiment.

Problem 2. Go back to the data of Experiment B-1 and use it to calculate values of K for each of Equations (5), (6), and (7). How do these values of K compare with each other?

We can now rewrite Equation (4) in a simpler form which may be easier to understand:

$$f = K\sqrt{\frac{T}{mL}} \qquad (8)$$

If the tension is increased, it increases the restoring forces that pull pieces of the string back and forth. This in turn causes the string to vibrate more rapidly. From Equation (8) you now know *how much* the frequency increases with increasing tension.

If the mass of the vibrating string is large, the string vibrates more slowly. Now you know *how much* the frequency decreases with increasing string mass.

The closer together the nodes of a simple vibration pattern are, the higher is the frequency of oscillation of the antinodes between the nodes. You observed this directly in Experiment A-2. Now you know quantitatively how frequency varies as the length of the vibrating string changes. But we have not yet explained why this happens.

Question 12. When you tighten a string on a guitar, which of the variables T, m, and L do you change? When you hold a string against a fret, which variables do you change?

Problem 3. Suppose that you decide to replace steel guitar strings with nylon strings on a particular guitar. The strings are to be tuned to the usual frequencies. If the mass per unit length of each nylon string is one-fourth that of the corresponding steel string, must the tension in the nylon strings be greater or smaller? By what factor must the tension be changed?

LOUDNESS, AMPLITUDE, AND FREQUENCY

By now you should understand how the guitar produces sound. The guitar strings, the sound board, and the air in the sound box vibrate. The air outside the sound box is, in turn, set into vibration. This sound is transmitted through the air to the ear in the form of sound waves. But what happens at the ear? We do not wish to delve deeply into the workings of this amazingly sensitive and complex organ. We shall just note that the sound wave travels through the ear canal to the eardrum, causing it to vibrate with the same frequency as the sound wave. This vibration is passed on to the small bones in the middle ear and eventually to the fluid and nerve endings in the inner ear. (See Figure 23). Here the vibrations are transformed into electrical nerve impulses which travel to the brain where they are interpreted as sound.

The loudness of the sound sensation caused by a vibrating structure, such as a tuning fork, depends primarily on the sound-wave amplitude. However, since your ear is not equally sensitive to all frequencies, the loudness is affected by the frequency.

FREQUENCY RESPONSE OF THE GUITAR

In Section A we discussed the fact that structures such as the sound board of a guitar or the air inside the sound box vibrate naturally at certain frequencies. These "natural frequencies" depend on the size, shape, elasticity, and mass of the vibrating structures. Such structures can be driven so that they vibrate at other frequencies. However, the amplitude of vibration of the structure is relatively small when driven at frequencies that are different from a frequency at which the structure vibrates naturally. If a structure is driven at or near one of its natural frequencies, the amplitude of response is large. This large response at the natural frequencies is called *resonance,* and the frequencies at which it occurs are called *resonant frequencies.*

To learn more about the loudness properties of sound you should now do Experiment B-2.

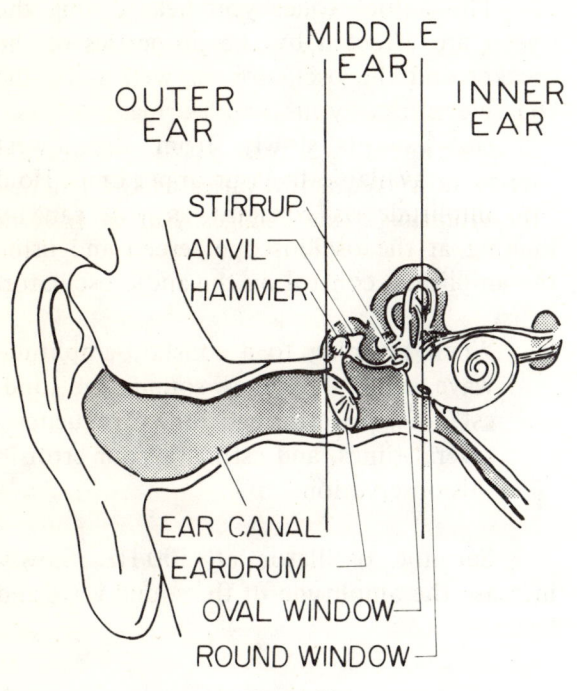

Figure 23.

EXPERIMENT B-2. Loudness and Resonance

PART I: Loudness

For this part of the experiment you are provided with an audio oscillator connected to a speaker. The audio oscillator will provide tones with frequencies you can vary over a wide range, including most of the audible range. You can also adjust the amplitude (and therefore the loudness) of the tones.

You can "see" the tones by using a microphone, which converts the sound waves into an electrical signal. The electrical signal from the microphone may be fed into an oscilloscope and displayed on the oscilloscope screen. The frequency of the wave on the oscilloscope screen is the same as that of the sound wave, and the amplitude is proportional to the sound-wave amplitude. This arrangement is shown in Figure 24.

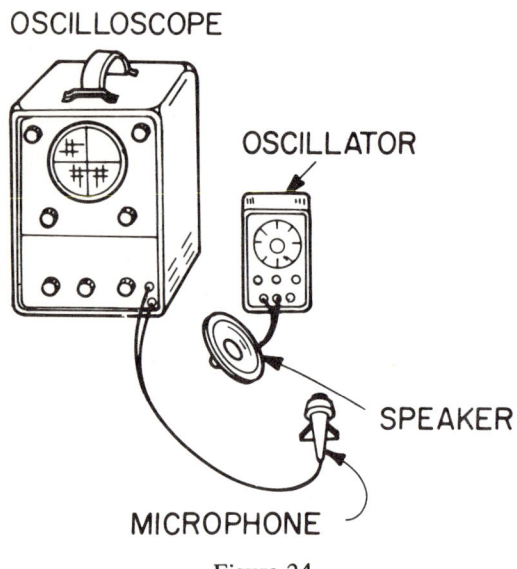

Figure 24.

Turn on the system and change the frequency of the audio oscillator slowly from its lowest frequency to 20 kHz while listening to the speaker. This operation is called a frequency *sweep*. Maintain a fixed position near the microphone as you listen to the frequency sweep.

1. How does the pitch of the sound change as the oscillator frequency increases?

The actual tones you hear during the sweep are affected by the properties of the speaker and the oscillator, as well as by the characteristics of your own hearing.

Now sweep slowly from the lowest frequency available to your upper limit. Hold the amplitude as constant as you can by looking at the oscilloscope screen and using the amplitude control on the audio oscillator.

2. When listening to a constant-amplitude wave, what frequency sounds the loudest? Sweep through the frequencies several times, and estimate your error in this observation.

Set the oscillator at 500 Hz. Slowly increase the amplitude of the sound wave and listen.

3. Does the loudness depend on sound-wave amplitude at this single frequency? Does the same effect hold for other frequencies? How does the pitch depend on sound-wave amplitude?

Set the oscillator to 3 kHz. Connect one terminal of a decade resistor to one of the terminals of the oscillator. Then to the other end of the resistor, connect one terminal of a push-button switch. Finally, connect the other terminal of the switch to the other terminal of the oscillator. With this arrangement you can change the amplitude of the signal, and thus the intensity of sound entering your ear, by pushing the button of the switch. This is because some of the electrical energy of the oscillator is consumed in the decade resistor. You can get different amplitudes just by adjusting the decade resistance. Although this is a crude method of attenuating the signal, over the range of amplitudes used, there will be no damage either to the oscillator or to the resistor. Also, there will be no appreciable distortion.

Set the decade resistor at 1 ohm and turn on the oscillator. Open and close the switch. (It is best to have someone else opening and closing the switch while you do *not* watch.) Set the amplitude as low as possible, but still loud enough that you can clearly hear the tones. Again, be sure that you remain at the same position near the microphone as you listen to the tones.

4. Can you detect a difference in loudness when the switch is opened and closed?

While the switch is being alternately opened and closed, adjust the decade resistor until you can just detect a difference in loudness; that is, to the point where, if you increase the resistance further, you no longer can hear any difference in loudness.

5. Record the amplitudes from the oscilloscope for the open and closed positions of the switch. (We assume that the amplitude of the input signal to the speaker is proportional to the amplitude of the sound which you hear.)

6. Square each of the two values of amplitude, and record these as measures of intensity. The *intensity,* defined as the energy carried by the wave per unit area per unit time, is proportional to the square of the amplitude.

To check the results of step 5, you may want to start with the highest decade resistance, where you cannot hear a loudness difference, and, as the switch is opened and closed, adjust the resistance down until you can first hear a difference in loudness.

Next set the amplitude of the oscillator to a high value where the tone is loud but not uncomfortable or too loud for others in the room.

7. Repeat steps 4 to 6 for this louder sound.

8. How does the intensity *difference* required for the smallest perceptible increase in loudness for the low-level sound at 3 kHz compare with the difference of intensities required for the louder sound at the same frequency?

9. How does the intensity *ratio* for the low-level sounds compare with the ratio for the louder sounds?

10. Set the oscillator to 400 Hz and repeat steps 4, 5, and 6. (Remember to set the initial amplitude to the lowest setting where you can still hear clearly the tones.)

11. How does the intensity *difference* required for the smallest perceptible increase in loudness for 3 kHz compare with the intensity difference at 400 Hz?

12. How does the *ratio* of the two intensities at 3 kHz compare with the ratio at 400 Hz?

PART II: Resonance

The equipment needed is a tuned guitar, an audio oscillator, a small speaker, and an intensity-level meter (called a decibel or dB meter; decibels will be defined later in this section of the module).

Set the guitar, neck up and face out, in a soft chair or on other material capable of absorbing stray sound. Place the speaker about six inches away from the sound hole and aimed at the hole. Connect the audio oscillator to the speaker. Hold the dB meter about two or three feet away from the guitar as shown in Figure 25.

Now turn on the oscillator and vary the frequency of its output, starting at low values and sweeping slowly across the entire range of hearing. Observe the readings of the dB meter as you make this sweep.

A frequency which sounds louder than others, and for which the dB meter reading is a maximum value, is called a *resonance* or a *resonant frequency.*

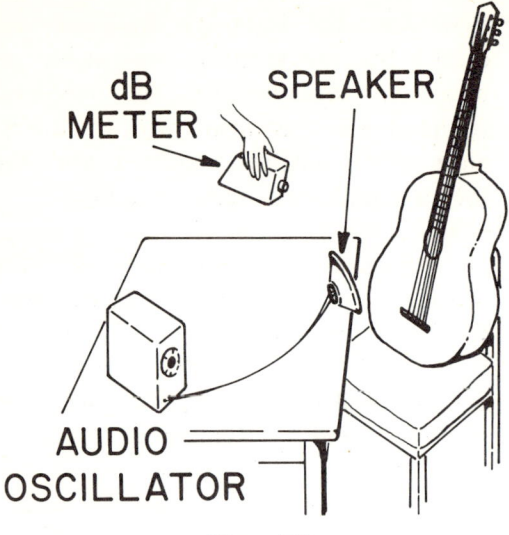

Figure 25.

1. How many resonances do you observe?

What is the source of these resonances? Is it the speaker? Is it the guitar body? Is it the strings on the guitar? Actually, all three of these vibrating objects are involved. The strings have their strongest resonances at their fundamental frequencies. The guitar body has resonances at frequencies which depend on its size and the way it is constructed.

Look at one of the string resonances. To do so set the oscillator at 294 Hz; then slowly adjust the tuning peg of the 4th string (the D string), first to loosen the string slightly, then to tighten it again. Repeat this until you feel the string resonate. You can do this by touching the string very lightly. If you place the guitar in a horizontal position, with the speaker directed downward into the sound hole, you can use a more sensitive method. Tear a tiny strip of paper, 1/8-in wide and 1-in long. Fold it and hang it over the string at the sound hole. When the resonance occurs, you can see the paper bounce up and down.

Adjust the other five strings by tuning them to the D string. Tune these strings by the standard method used by guitarists (your instructor will explain this simple technique if you don't already know it).

2. Now, use the strip of paper as a detector and slowly adjust the oscillator frequency upward, starting at about 150 Hz. In this way find the resonant frequency of each of the other five strings. (You may have difficulty seeing these resonances for the first and second strings, but do try.) Record these frequencies on the worksheet.

Now, remove the guitar and use the dB meter to look for resonances in the speaker. The oscillator produces a fairly "flat" (same sized) signal over the range of frequencies you are listening to. Therefore, if the dB meter shows a maximum reading at some frequency, it is a resonant frequency of the speaker.

3. What is the most pronounced resonant frequency of your speaker?

Now direct the speaker back into the sound hole of the guitar and look for another resonance. Be sure that it is *not* the speaker resonance. Where you find a resonance, feel the strings to see if any one of them shows a strong resonance. If so, you may be at a string resonance.

If several of the strings show some vibration, this could be because the guitar body is resonating and transferring some vibrations to the strings. When you are sure the resonance is for neither the speaker nor for any of the strings, it must be in the guitar body itself.

When you find a body resonance you can study it in more detail. First, put the dB meter in a fixed location, about two feet from the guitar, and vary the frequency of the oscillator from about 100 Hz below the resonant frequency to about 100 Hz above the resonant frequency.

4. Record the readings of the dB meter at several different frequencies in this range, say about every 20 Hz. Plot a graph of the intensity level recorded on the dB meter versus frequency of the signal. This graph is called a *response curve*.

5. At resonance, can you feel the vibration of the sound board of the guitar? One way to look for these vibrations is to

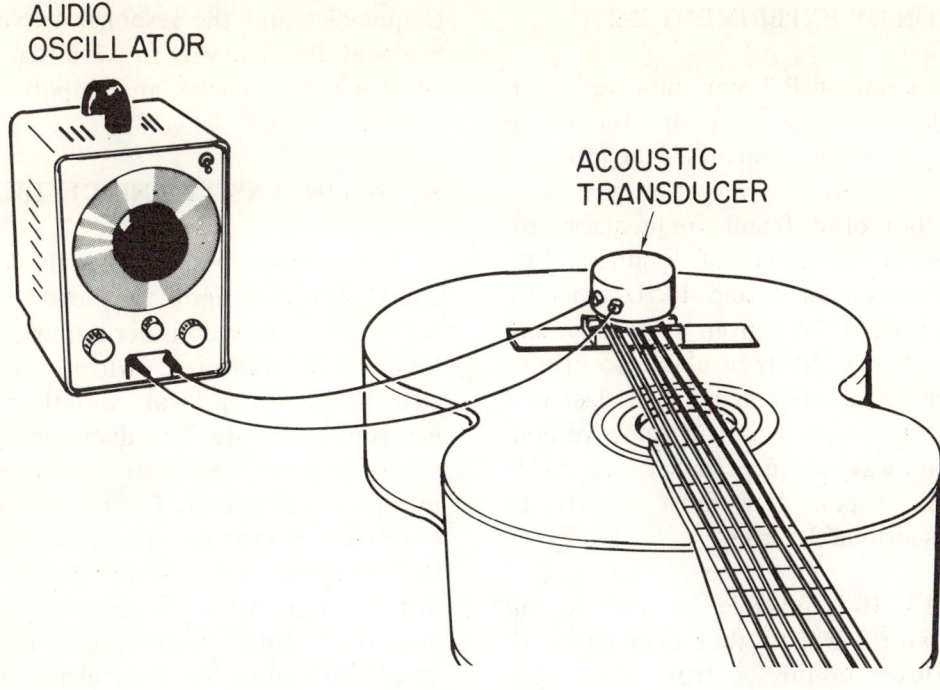

Figure 26.

sprinkle cork dust over the sound board of the guitar when it is horizontal (with the speaker directed into the sound hole from above).

If the sound board and other parts of the surface of the guitar body were not vibrating much at the resonance you found, you may have found the so-called *Helmholtz resonance*. The Helmholtz resonance results from vibrations of the air inside the box, and is the same kind of resonance you get from blowing gently across the top of a soft drink bottle. You can check for a Helmholtz resonance by covering the sound hole with a piece of cardboard. If the sound of the resonance greatly decreases in loudness, then you have found a Helmholtz resonance.

If the sound board does vibrate a lot, then the resonance is one of the natural modes of vibration of the sound board. If this is the resonance you observed, see if you can find the frequency for the Helmholtz resonance.

Now go back to one of the string resonances.

6. How much must you change the frequency from resonance before the string ceases to vibrate a noticeable amount?

7. Which resonance has the "widest" response curve, that for a string, or for the guitar body or sound board?

To better observe the resonances of the guitar body and strings, you have been provided an *acoustic transducer*. This device is the same as a speaker, but with the speaker cone replaced by a screw. This transducer can be touched to objects to make them vibrate. Connect the oscillator output to the transducer as shown in Figure 26. Set the oscillator at 294 Hz and hold the transducer in the position shown in the figure so that the screw touches the 4th string (D) about 1/8 in from the bridge. Then slowly adjust the oscillator frequency above and below 294 Hz until the string begins to vibrate. Again, if you have trouble seeing the oscillations, place a small folded piece of paper over the string at the sound hole.

8. Using the same method, locate the resonant frequencies for the other strings and for the guitar body. Record these results.

DISCUSSION OF EXPERIMENT B-2

In Experiment B-2 you observed that, for a single frequency, increasing the amplitude of the sound wave increases the loudness of the sound. Sound waves of the same amplitude but of different frequencies produce different sensations of loudness. Frequencies between 3 kHz and 4 kHz generally sound the loudest for a given amplitude, and frequencies below 20 Hz or above 15 kHz to 20 kHz are not audible at all. This last fact means that the typical human ear is responsive to sound-wave frequencies between 20 Hz and 20 kHz at best. It is most sensitive to frequencies around 3.5 kHz.

Question 13. It is possible for you to call your dog with a whistle that is inaudible to humans. How can this be true? Make your answer in terms of concepts such as *frequency* and *sensitivity*.

In Part II of Experiment B-2, you found resonance for which there was little vibration of the sound board. This particular resonance is called the *Helmholtz resonance,* and it is named for a nineteenth century German scientist. The Helmholtz resonance involves pressure oscillations throughout the sound box at a frequency determined by the properties of the air within the sound box. This kind of resonance also causes the musical sound you hear when you blow across the mouth of an empty bottle. You detected other resonances. These resonances are associated with sound-board vibrations.

Most of the sound coming from a guitar is caused by vibrations of the wooden sound board and the air in the sound box. These vibrations are excited by the vibrating strings. Since the guitarist controls the frequencies of the vibrating strings, a sound of any desired pitch can be obtained. However, the intensity of this sound would be small if the string vibrations were not coupled to the bridge, and thus in turn to the sound board and the air cavity. The response of these structures is large at certain resonant frequencies. Fortunately there are several of these resonant frequencies and the response curve for each resonant frequency is so wide that sounds at all string frequencies are amplified to some extent.

SOUND INTENSITY AND DECIBELS

You have observed how the loudness of sound depends upon frequency and amplitude. Let us now consider in more detail how loudness is measured. When a sound wave strikes your ear, pressure variations cause the eardrum to vibrate. This disturbance produces signals that are transmitted to your brain and interpreted as sound. The ear is an extremely sensitive detector of sound. For soft sounds, the eardrum may move through a total distance (amplitude) smaller than the diameter of a single atom (10^{-10} m) and still detect the sound. The total change in pressure associated with such a sound may be only one ten-billionth (10^{-10}) of atmospheric pressure. If the eye could respond to an amount of light energy equal to the smallest amount of sound energy audible to the ear, one could see 50 W of visible light from 3000 miles away. (This would require about a 500-W bulb, since only about 10% of the radiation from a light bulb is visible.) On the other hand, the ear also responds to sounds a trillion (10^{12}) times more powerful than the minimum audible sound.

Question 14. a. When you nod your head, the slight altitude change produces a pressure change of about 10^{-10} that of atmospheric pressure. If the ear is so sensitive to changes in pressure, why do you not hear a tone when you nod your head up and down rapidly?

b. If a meter stick which is tightly clamped to a table top so that most of its length protrudes over the edge is set into vibration, you do not hear the vibrations. But, if it is clamped with just a short length protruding over the table edge, the vibrations are clearly audible. Why?

Starting with the plucking of the guitar string, and in all other instances of sound production that you have observed, the "size" or amplitude of the disturbance is somehow related to the sensation of loudness. It turns out that the relation is more subtle than one might first guess. For example, doubling the amplitude does *not* give twice the initial loudness. We shall now see what measures of loudness are appropriate and useful.

The frequency of a sound wave is used as a measure of pitch. The amplitude of a sound wave is related to its loudness, but amplitude alone is not a convenient measure of loudness. The energy carried by the wave is proportional to the *square* of the amplitude, but even this is not a convenient measure of loudness. To make things more complicated, the human ear does not respond linearly to the wave energy; a wave carrying twice as much energy does not sound twice as loud. The energy carried by a wave is measured by the quantity called *intensity I*. The intensity is defined as the amount of energy which passes through a unit of area (perpendicular to the direction of propagation of the wave) in one second of time. Thus if a total amount of energy E passes through an area A in a time interval Δt*, we define the energy per unit area per unit time by

$$I = \frac{E}{\Delta t \cdot A} \quad (9)$$

(Definition of Intensity)

Energy delivered per unit time is called *power*. You are familiar with some common units of power in such uses as a 40-*watt* light bulb or a 300-*horsepower* engine. Mathematically, we can write

$$P = E/\Delta t \quad (10)$$

(Definition of Power)

so that

$$I = P/A \quad (11)$$

Intensity has units of watts per square meter (W/m^2).

*The Greek letter *delta* (Δ) is often used to indicate a change or difference in a quantity. Here it indicates an elapsed time interval.

Example Problem. In normal conversation, the sound at your ear has an intensity of about 10^{-7} W/m^2. If the area of the eardrum is 5×10^{-5} m^2, what sound power (energy per unit time) passes into the ear?

Solution. Given in the problem are

$$I = 10^{-7} \text{ W/m}^2$$

and

$$A = 5 \times 10^{-5} \text{ m}^2$$

The equation $I = P/A$ must be rearranged in order to solve for the power. If both sides of that equation are multiplied by A, we obtain

$$P = IA$$

Substituting the given values into this equation,

$$P = 10^{-7} \text{ W/m}^2$$
$$\times 5 \times 10^{-5} \text{ m}^2$$

Performing the power of ten multiplication and gathering units,

$$P = 5 \times 10^{-12} \text{ (W/m}^2\text{)} \times \text{m}^2$$

or

$$P = 5 \times 10^{-12} \text{ W}$$

Question 15. If the power of sound entering your ear during normal conversation is 5×10^{-12} W, why would you need a stereo system amplifier rated at 10 W to 100 W?

Problem 4. The light from a 100-W light bulb has an intensity of 0.4 W/m^2 at a distance of 4 m from the bulb. If the pupil of your eye has an area about equal to the area of the eardrum (5×10^{-5} m^2) and is located 4 m away from the bulb, how much power is received in your eye from this light source?

Problem 5. Traffic on a busy city street creates a sound intensity of 10^{-4} W/m^2. If the eardrum has an area of 5×10^{-5} m^2, what power impinges on the eardrum?

The human ear can distinguish between two frequencies that differ by as little as 0.3% in the range from 500 Hz to 4 kHz and much less than this for musical tones made up of harmonics. For example, at 1000 Hz, the frequency must increase to about 1003 Hz before your ear can detect a change. At 3000 Hz, the frequency must increase to about 3009 Hz for your ear to detect the change. The ear does not detect a fixed difference of frequencies but, instead, detects a change given approximately by the *ratios* of the two frequencies. That is,

$$f_2/f_1 \cong 1.003$$

where f_1 is the original frequency and f_2 is the value of frequency just enough higher than f_1 that your ear can perceive a pitch change under ideal conditions.

As you observed in Part I of Experiment B-2, the ear responds to loudness in a similar way. Suppose you are listening to a 1-kHz tone and the intensity at your ear is 1.62×10^{-7} W/m^2. In order to hear a change in loudness, the intensity at your ear must increase to about 1.80×10^{-7} W/m^2. These two intensities have a *difference* of $(1.80 \times 10^{-7} - 1.62 \times 10^{-7})$ W/m^2. One might expect that increasing the intensity by another 0.18×10^{-7} W/m^2 would again give a detectable change in loudness. But in fact it doesn't! The intensity must be increased from 1.80×10^{-7} to 2.00×10^{-7} W/m^2, a difference of 0.20×10^{-7} W/m^2, to be noticeable. However, the ratios 1.80/1.62 and 2.00/1.80 are both equal to 1.11. Your ear does not detect a fixed *difference* of intensities. Instead, it detects an intensity change given by fixed *ratios* of the two intensities. That is,

$$I_2/I_1 \cong 1.11$$

where I_1 is the original intensity and I_2 is the value of intensity just enough greater than I_1 that your ear perceives a change in loudness.

As you probably found in Experiment B-2, the ratio from one intensity to the next perceptible intensity is not exactly 1.11. Indeed, this ratio may change considerably for different frequencies, and for different levels of loudness. In an acoustical laboratory, the ratio 1.11 can be obtained for a rather wide range of frequencies and loudness levels. However, for the ordinary person listening in a room, the ratio may be closer to 1.25.

At very low and very high sound levels, and for frequencies which are at the extremes of what people can hear, the ratios for perceptible changes in loudness are greatly different from the usual range of 1.11 to 1.25. For example, at 35 Hz, a barely audible frequency, the ratio is about 8. For a very loud tone at 1000 Hz, the ratio is about 1.06. For a barely audible tone at 1000 Hz, the ratio is 2.

Since the ratios are not really very constant, you may well ask why the ratio "law" is significant. The important thing to realize is that the *differences* in sound intensities for perceptible changes vary enormously, while the ratios of these same intensities are nearly constant. This means that a ratio is more nearly correct (instead of a difference) in describing how our ears respond to intensity changes. For example, at 1000 Hz, a barely audible tone must have an intensity difference of about 3×10^{-12} W/m^2 to be perceptible. For a loud tone at that same frequency, the difference must be 6×10^{-4} W/m^2. The latter difference is some 200 million times greater than the intensity difference required for a barely audible tone. Whereas, the ratio for one case is only 3 times larger than that for the other.

Decibels

In other words, before your ear can detect a loudness change, the intensity of sound must increase by between 11% and 25%. This physiological response to sound is one example of the way your senses respond to all stimuli. For example, you respond in a similar way to pressure differences on your skin and to changes of light intensity in your eyes. For perceiving changes in touching pressure, the ratio is about 1.03. For changes in light intensities, the ratio is about 1.008. In terms of percentages, the pressure must change by about 3% to be detectable, light intensity must change by about 0.8%. The general principle involved here is called the *Weber-Fechner Law: A change in the strength of a stimulus necessary to produce a perceptible difference in sensation is proportional to the intensity of the stimulus already acting.*

The numbers and percentages quoted above are determined by sampling the responses of large numbers of people. Thus, they represent average, or typical, responses. Because the ratios do change so much over the range of hearing, and because the subjective sensation of loudness is far more complicated than we have indicated, the Weber-Fechner "law" *cannot* be considered a scientific law in any strict sense. Still it is a good approximation, and it makes the important point that the ratios rather than differences are more nearly constant.

Question 16. In your own words present an argument showing that the Weber-Fechner Law does indeed describe the idea illustrated in the preceding numerical examples.

Because your ear responds to sound intensity only in steps based on ratios, let us consider these steps. Suppose we let I_0 stand for the intensity of a tone which is barely audible. Then the next higher intensity of that tone which is discernible at a different intensity is 1.11 times I_0, or $1.11\, I_0$. The next level of intensity which is perceptibly louder is 1.11 times this latter intensity, or $1.11\,(1.11)\, I_0 = (1.11)^2 I_0$. This procedure can be repeated for each discernible increase in intensity level. We can continue in this way, constructing Table I.

Table I.

Intensity	Number of Discernible Steps above Audible Level
$I_0 = (1.11)^0\, I_0$	0
$I_1 = 1.11\, I_0 = (1.11)^1\, I_0$	1
$I_2 = 1.11\, I_1 = (1.11)^2\, I_0$	2
$I_3 = 1.11\, I_2 = (1.11)^3\, I_0$	3
$I_4 = 1.11\, I_3 = (1.11)^4\, I_0$	4
⋮	⋮
$I_n = 1.11\, I_{n-1} = (1.11)^n\, I_0$	n

Problem 6. Fill in Table I for the cases of five and six discernible steps above audible level.

The important thing about this table is that, for a given number of discernible steps above the audible level, the intensity is given by an expression which contains the step number as an exponent. For example,

$$I_4 = (1.11)^4 I_0$$

for step number 4. Expressed as a ratio, we have

$$I_4/I_0 = (1.11)^4$$

or, for the nth step,

$$I_n/I_0 = (1.11)^n \tag{12}$$

If we take the logarithm of each side of this equation, we get

$$\log(I_n/I_0) = \log(1.11)^n$$
$$= n \log(1.11)$$

where all the logarithms are to the base 10.

Solving for n, we get

$$n = \frac{1}{\log 1.11} \log(I_n/I_0)$$

or

$$n \cong 22 \log(I_n/I_0) \tag{13}$$

These step numbers are "equally spaced" as far as our ability to detect loudness changes for a particular frequency tone is concerned. These step numbers vary with the *logarithm* of the intensity ratio, and not with the ratio itself. For this reason, we say that our ears "respond logarithmically" rather than directly to changes of sound intensity.

In technical work it is conventional to use a unit of intensity level which is different in size from the discernible step level our ears perceive. This unit, called the *bel*, is named after Alexander Graham Bell. The intensity level ($I.L.$) in bels is given by

$$I.L. = \log(I/I_0) \tag{14}$$
(Definition of sound intensity level in bels)

From Equations (13) and (14), the number of discernible intensity steps between the two given levels, I_0 and I, is related to $I.L.$ by the expression

$$n \approx 22 \, I.L. \tag{15}$$

Thus, there are about 22 discernible intensity steps in one bel. One bel is therefore a very large unit (like using pounds when ounces would be more appropriate). A smaller unit is much more convenient. This more commonly used unit is one-tenth of a bel and is called a *decibel* (dB). There are 10 dB in 1 bel:

$$I.L. \text{ (in dB)} = 10 \log(I/I_0) \tag{16}$$

If a decibel is one-tenth the size of a bel, the decibel is about the size of 2.2 discernible intensity steps, or

$$n \approx 2.2 \, I.L. \text{ (in dB)} \tag{17}$$

The number of discernible steps of sound level is not an exact number. There is considerable variation according to frequency, loudness, and the individual. The ratio 1.11 is only approximate, so that the constants in Equations (12), (13), (15), and (17) are only approximate. It is the *approximately* logarithmic response of our sense of hearing which is important. Thus, Equations (14) and (16), which are exact definitions, serve scientific purposes, whereas the less exact equations based on the approximate, experimentally determined ratio 1.11 merely serve to help us understand *why* we use a logarithmic form for intensity level. To provide a common base or reference level relative to which intensities are measured, the numerical value 10^{-12} W/m^2 is chosen as the minimum audible sound intensity. Thus

$$I_0 = 10^{-12} \text{ W/m}^2 \qquad (18)$$
(Accepted minimum audible sound intensity)

Example Problem. The sound level for normal conversation was given earlier as 10^{-7} W/m². What is this intensity level in bels? What is it in decibels? About how many discernible loudness levels is normal conversation above a barely audible sound?

Solution. Given in the problem is

$$I = 10^{-7} \text{ W/m}^2$$

The minimum audible sound level is

$$I_0 = 10^{-12} \text{ W/m}^2$$

Substituting these values into Equation (14), we have

$$I.L. = \log\left(\frac{10^{-7} \text{ W/m}^2}{10^{-12} \text{ W/m}^2}\right)$$

or

$$\text{Intensity level} = \log(10^5)$$

The base ten logarithm of 10^5 is 5; thus

$$I.L. = 5 \text{ bels}$$

Since there are 10 dB in every bel, the intensity level for conversation is 50 dB. (From this point on, the word "decibel" will be replaced by the symbol "dB.")

From Equation (17) we see that there are about 2.2 discernible sound level steps in each dB; therefore there are about 2.2 × 50, or 110 discernible sound level steps from a barely audible sound to the sound level of normal conversation.

Example Problem. The sound level in a factory is 98 dB. What intensity is this in W/m²?

Solution. The intensity level is 98 dB, and the reference intensity $I_0 = 10^{-12}$ W/m². The intensity level must be expressed in bels and Equation (14) may be used to solve for I.

Taking the antilog of both sides of Equation (14) gives

$$\text{Antilog } (I.L. \text{ in bels}) = I/I_0$$

Multiplying both sides by I_0 then gives

$$I = [\text{antilog } (I.L. \text{ in bels})] \times I_0$$

The intensity level in bels for this example is 9.8. Substituting into the rearranged equation gives

$$I = [\text{antilog } (9.8)] \times 10^{-12} \text{ W/m}^2$$

From a table of common logarithms,

$$I = [6.3 \times 10^9] \times 10^{-12} \text{ W/m}^2$$

or

$$I = 6.3 \times 10^{-3} \text{ W/m}^2$$

Problem 7. Hearing becomes painful at a sound level of about 1 W/m². Calculate the sound intensity level in bels for this sound level. What is the intensity level in dB? About how many discernible loudness levels above a barely audible sound is this threshold of pain?

Problem 8. The sound made by the feet of a small dog running across the floor has an intensity of 3.2×10^{-11} W/m². What is the intensity level in bels? What is it in dB?

Problem 9. The sound of an electric motor has an intensity level of 55 dB. Calculate the intensity of this sound in W/m².

Problem 10. For each intensity level in the following table, calculate the intensity in W/m². Express each numerical value as a number times 10^{-12}.

HEARING RESPONSE

Figure 27 is a graph that shows the hearing ability of the population as a whole. Each curve on the graph is labeled with a number which gives the percentage of the population who can hear that intensity and

Table II. Intensity Levels of Some Familiar Sounds

Threshold of hearing	0 dB	Very heavy traffic in a large city	80 dB
Gentle rustle of leaves	10 dB	Subway station with express passing, New York	95 dB
Whisper at 4 ft	20 dB	Power lawnmower	110 dB
Quiet suburban street	30 dB	Steel plate hammered by four men, 2 ft away	112 dB
Quietest time at night, center of a large city	40 dB	Sounds of the firing of a .357 Magnum pistol, 3 ft away	120 dB
Conversation at 12 ft	50 dB		
Busy traffic in a large city	65 dB		

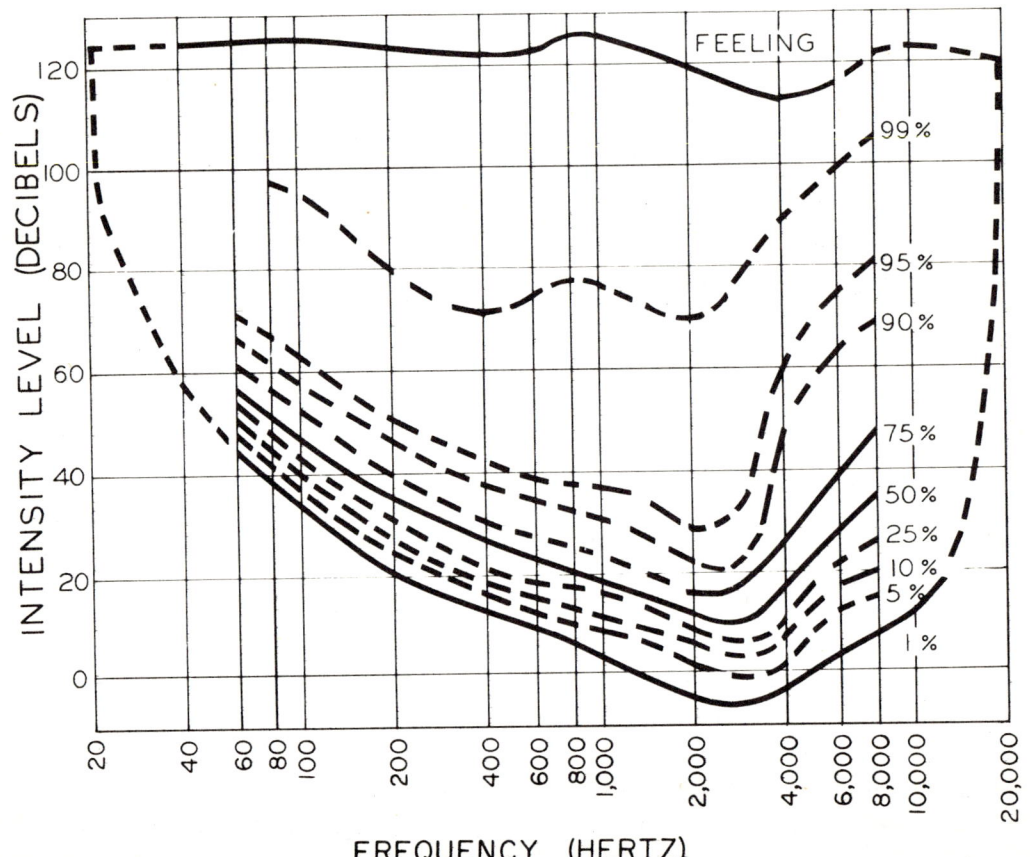

Figure 27. Hearing ability of the population.

frequency. The top curve (called the *threshold of feeling* curve) is the highest intensity perceived as sound. Higher intensities are perceived as pain. The bottom (1%) curve is usually taken as "normal" hearing. It is assumed that, if 1% of the population hear tones at these intensities, that is "normal" hearing. The other 99% of the population, for one reason or another, has lost this hearing ability. Even for the 1% curve, the threshold is not 0 dB at all frequencies. For example, at 200 Hz the intensity level must be 20 dB before the sound is perceived by those with normal hearing. The 50% curve is indicative of "average" hearing, because half the population hears better and half hears worse. Note that "average" hearing is not "normal": at 200 Hz the intensity level must be about 35 dB before 50% of the population hears it.

Question 17. What frequency sound is heard by the largest number of people at a level of 10 dB? About what fraction of the population can hear it?

Question 18. What fraction of the population can hear 100 Hz at 20 dB? At 60 dB?

Question 19. Between about 1300 Hz and 5000 Hz, the 1% curve in Figure 27 is below 0. What does this mean? Can it be correct? Explain your answer carefully.

INTENSITY LEVEL IN PHONS

Figure 27 indicates that the ear is not equally sensitive to all frequencies. Although the decibel scale corresponds roughly to the sensation of loudness at a single frequency, it cannot be used to compare the relative loudness of two sounds at different frequencies. For the latter purpose a new scale of loudness, in units called *phons*, is used. This scale of loudness is based on hearing response at a frequency of 1000 Hz. The *loudness level* (in phons) of any tone is defined as the intensity level (in decibels) of a 1-kHz tone of equal loudness. Thus, a tone of frequency 1 kHz and intensity level 40 dB has a loudness level of 40 phons. The loudness of other frequencies is determined by experiments. While one cannot accurately judge changes in loudness, one can tell when two sounds are equally loud. In the experiments, people were asked to judge when sounds of different frequencies were as loud as a 1-kHz sound. In this way, *curves of equal loudness* were generated for various loudness levels. The results are summarized in Figure 28, where each equal-loudness curve is labeled by its phon value. The phon has only limited usefulness. It is actually used to arrive at another measure of loudness, the subjective loudness expressed in *sones*. We will not consider sones in this module.

Example Problem. What is the loudness level in phons of a 100-Hz tone at an intensity level of 60 dB?

Solution. Since loudness is a subjective quantity, and the data has been presented here in the form of a graph, the solution is obtained by referring to Figure 28. The intersection of the frequency (horizontal) axis with the intensity-level (vertical) axis is read in relationship to the loudness curves (labeled in phons). In this example the 100-Hz line intersects the 60-dB intensity level line at a point about halfway between the 30- and 40-phon curves. The loudness is then about 35 phons.

Example Problem. What intensity level must a 10-kHz tone have in order to sound just as loud as a 1-kHz tone with an intensity level of 40 dB?

Solution. We know that the loudness level for the 1-kHz sound is 40 phons. We then refer to the graph of Figure 28 and follow the 40-phon curve until it intersects the 10-kHz line. This intersection corresponds to an intensity level which may be read on the vertical axis. In this case that intersection corresponds to an intensity level of about 52 dB.

Problem 11. What is the loudness level in phons of a 400-Hz tone at an intensity level of 20 dB?

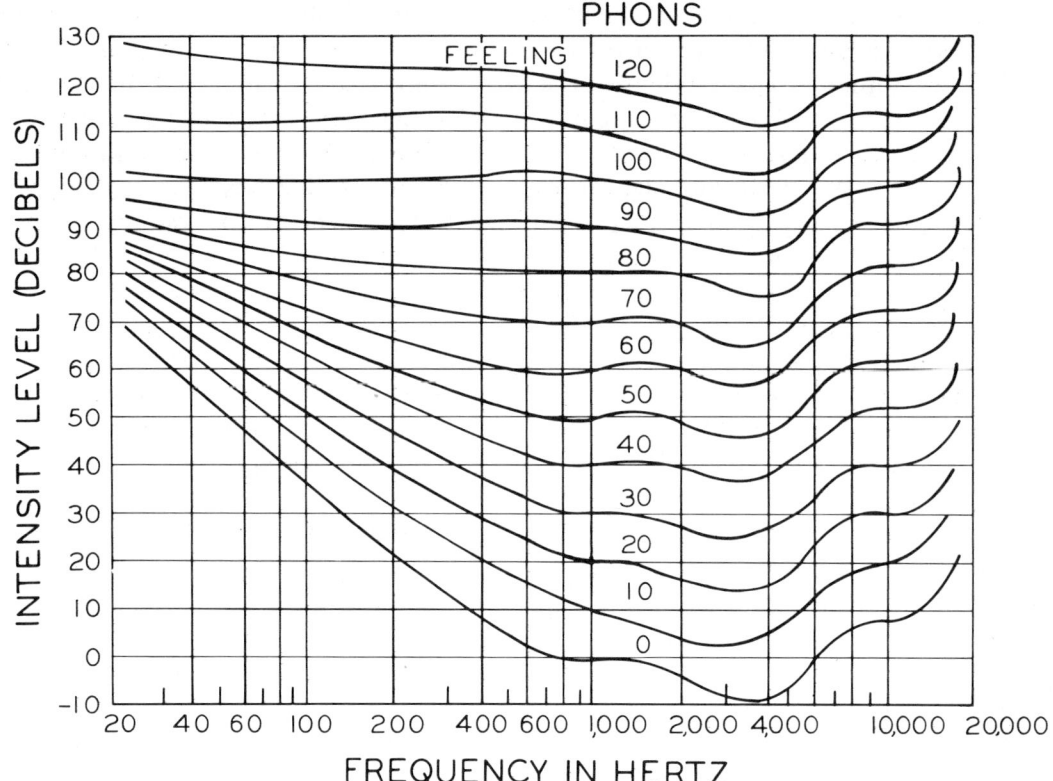

Figure 28. Curves of equal loudness.

Problem 12. What intensity level must a 40-Hz tone have in order to sound just as loud as a 1-kHz tone with an intensity level of 60 dB?

Question 20. Is there any frequency other than 1 kHz at which a sound with an intensity level of 40 dB has a loudness of 40 phons? If so, at what frequency does this occur?

SUMMARY

In this section you have discovered the empirical relationship between the fundamental frequency f of a string and properties of the string, the tension T, mass m, and length L. This relation can be written as

$$f = K\sqrt{T/mL}$$

where K is a constant of proportionality.

You have also found that the frequencies to which the human ear responds range from about 20 Hz to 20 kHz, with the best response falling between 3 and 4 kHz. Two amplitudes such that the ratio of their squares is a constant, regardless of what amplitude you begin with. Saying this differently, the ratio of intensities is constant.

The sound of a guitar depends on the resonances between the oscillations of the string and the possible oscillations of the air inside the guitar box or those of the sound board.

You learned definitions of sound-wave intensity and level. Intensity I is given by

$$I = P/A$$

where P is the power carried by the sound wave and A is the area over which that power is spread. Intensity level $I.L.$ is defined in terms of intensity I and the minimum audible intensity (arbitrarily chosen as $I_0 = 10^{-12}$ W/m^2):

$$I.L. = \log_{10}(I/I_0) \text{ in bels}$$

or

$$I.L. = 10 \log_{10} (I/I_0) \text{ in decibels (dB)}$$

Intensity level is a measure of loudness at a single frequency. The number of discernible steps n in loudness at a given frequency is given approximately by

$$n = 2.2 \, I.L. \text{ (in dB)}$$

Finally, you learned that the variation of loudness sensation with frequency is specified by graphical information, such as the whole population hearing curves of Figure 27 and the curves of equal loudness of Figure 28.

GOALS FOR SECTION C

The following goals state what you should be able to do after you have completed this section of the module. These goals should be studied carefully as you proceed through the module and as you prepare for the post-test. The example which follows each goal is a test item which fits the goal. When you can correctly respond to any item like the one given, you will know that you have met that goal. Answers appear immediately following these goals.

1. *Goal:* Understand how the superposition of two traveling waves produces a standing wave.

   Item: Sketch the resultant standing wave between B and C at the instant the two traveling waves are as shown in Figure 29.

   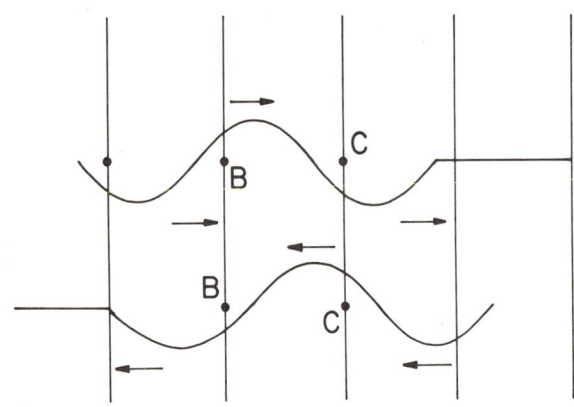

   Figure 29.

2. *Goal:* Know the relationship between the speed, wavelength, and frequency of a wave.

   Item: If the speed of sound in air is 340 m/s, what is the wavelength of a 440-Hz tone?

3. *Goal:* Know the general formula for the frequencies of the normal modes of oscillation of a string which is fixed at both ends.

   Item: What is the frequency of the third harmonic of a string 40 cm long whose mass is 20 g and which is under 64 N of tension?

4. *Goal:* Understand how the superposition of harmonics on a string produces any particular standing wave.

   Item: A guitar string is plucked at one-fourth the distance from bridge to nut. What harmonics are missing? Is the second harmonic very small in amplitude compared to the fundamental or not? (Use sketches to decide, if necessary.)

5. *Goal:* Understand the physical basis of musical intervals, consonance, and dissonance.

   Item: The three notes immediately above E are F, G, and A. Which of the three form consonant intervals with E and what are the intervals?

Answers to Items Accompanying Previous Goals

1. See Figure 30. Less than the full maximum displacement.

   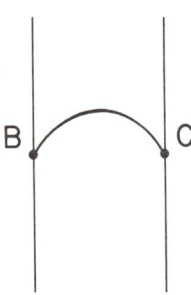

   Figure 30.

2. .773 m.

3. 134 Hz.

4. 4th, 8th, 12th, etc. missing. The amplitude of the second harmonic will be relatively large compared to its amplitude when the string is plucked at other points.

5. G is consonant, forming a minor third; A is consonant, forming a fourth; F is not consonant.

SECTION C

TRAVELING WAVES ON A STRING

You have discovered that some patterns of vibration on a guitar string have simple properties. Each pattern has nodes at both ends, where the string cannot move. The simplest pattern has amplitudes of oscillation that increase gradually from the ends to a maximum, called an antinode, at the midpoint. Other simple patterns are similar, except that additional nodes and antinodes appear. On strings, the nodes are always equally spaced. Associated with each pattern is frequency given by an equation that you discovered empirically in Experiment B-1:

$$f = K\sqrt{\frac{T}{mL}}$$

where T is the tension in the string, m and L are the mass and length of the portion that vibrates, and K is a constant that can be determined by measurement.

Why does the string behave this way? Can we propose a set of simple ideas about what goes on when the string vibrates such that the string equation above becomes a necessary result? Such a set of ideas is called a *model* or, if it encompasses a wide range of phenomena, a *theory*. If the equation derived from the theory agrees with the equation that was determined from experiment, then we regard the theory as tentatively correct. A good theory can also be used to predict the results of experiments not yet performed. If the results of these new experiments agree with the theoretical predictions, we gain more confidence in the theory. If they do not, we may either discard the theory and replace it with a new one, or modify it so that it predicts correctly the results of all the experiments we know about. Note that there is a big difference between facts and empirical relationships on the one hand, and models and theoretical equations on the other. The former are not matters of opinion and they are not subject to change, except for refinements that result from reductions in the errors of measurement. The latter are creations of human minds, and they must be modified whenever new evidence indicates that they are inconsistent with the facts.

What we now seek is a theory from which we can deduce the string equation as a necessary consequence. Experiment C-1 will help to suggest a suitable model for the behavior of vibrating strings.

EXPERIMENT C-1. Transverse Pulses on a Spring

The equipment needed is a long spring and a timer. You have already observed that there is similarity in the patterns of transverse vibrations on stretched springs and on guitar strings. Since the patterns on the spring are easier to see and the frequencies produced are lower, it requires less equipment to make measurements using a spring.

Stretch the spring between two points on a smooth floor; keep these endpoints fixed during the measurements described below. In this way, the spring will be similar to a guitar string fixed at each end. Near one fixed end pull a piece of the spring to one side and release it. Watch the pulse travel back and forth along the spring. Count the number of such traversals that you can follow clearly. Now repeat this procedure, but use a stopwatch to measure the time it takes for the pulse to complete as many trips back and forth as you can easily observe. (On a guitar string, a pulse like this moves much too fast to observe and time with a stopwatch.)

1. What is the time required for one complete round trip of the pulse? (A round trip is the motion of the pulse from its starting point down to the far end, back to the near end, and back to its starting point.)

With the spring raised up off the floor, push the spring gently to and fro sideways near one end so as to excite the simplest pattern of vibration. That is, the pattern is the one with nodes on the ends and an antinode at the midpoint. This situation is the same as when a guitar string vibrates at its fundamental frequency. When this pattern is well established, measure the time it takes to complete some predetermined number of complete oscillations. (A complete oscillation is the motion of the spring from a maximum displacement on one side to a maximum displacement on the other, then back to its original position.)

2. What is the time required for one complete oscillation? This time is called one *period of oscillation,* or more simply, one *period.*

3. What is the frequency of oscillation? How is the frequency related to one period?

4. How does the time required for a pulse to travel to the end of the spring and back compare with the time found in Question 2 for one complete oscillation?

Now with the spring back on the floor, create a single transverse pulse and watch it travel back and forth along the spring.

5. Describe how the shape of the pulse changes as it travels along. Would you say that shape change is rapid or slow?

6. Describe any change in amplitude (maximum displacement) that occurs as the pulse travels along. Would you say that the amplitude change during one trip down the spring is large or small?

7. After the pulse reflects from an end

    a. Is the pulse displacement on the same side of the spring as before reflection or on the opposite side? If the reflected pulse is on the same side, we say it is reflected in the same *phase* as the incident pulse. If it is on the opposite side, we say it is *inverted* relative to the incident pulse.

    b. Is the shape of the pulse much different from its original shape?

Now create a pulse with a small amplitude by striking the spring lightly with the edge of your hand near one end (a gentle karate chop). Time the back-and-forth motion as you did earlier. Next create a larger amplitude pulse of similar shape by striking the spring harder at the same place. Again time the back-and-forth motion.

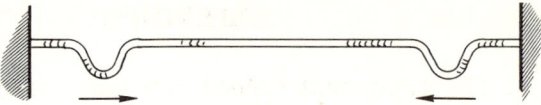

Figure 31.

8. Is the time required for one complete round trip nearly the same or quite different for two pulses of different amplitude?

Now create a broader pulse by quickly moving the end of the spring to one side, holding it there briefly, then moving it quickly back to its initial position. Measure the time it takes the pulse to complete a round trip.

9. Is the time required for one complete round trip nearly the same for a broad pulse as for a narrow one?

For the next part of the experiment, you will need the cooperation of at least two other persons. One person should count cadence and at the signal, GO, the other two should introduce pulses at opposite ends of the spring. The two pulses should be as similar as possible, and on the same side of the spring, as shown in Figure 31. This will take some practice. Watch what happens in the region where the two pulses overlap briefly. Also try to follow one of the pulses after it emerges from this overlap region. (If things happen so fast that you can't follow them, reduce the tension in the spring and the pulses will travel more slowly.)

10. Is the maximum displacement in the overlap region greater than, equal to, or less than the amplitude of either one of the original pulses? Do you think it is greater than, equal to, or less than the sum of the amplitudes of the two original pulses?

11. After a pulse emerges from the overlap region, is it essentially the same as or completely different than it was before it entered the overlap region?

Now repeat the previous observation, but with one important difference: one original pulse should be on one side of the spring, and the other pulse should be on the other side of the spring. This situation is shown in Figure 32. We say that the displacement on one side is *positive* and the displacement on the other side is *negative*.

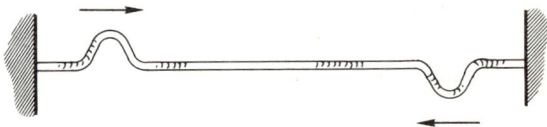

Figure 32.

12. When the centers of the two pulses reach the midpoint of the spring, is the displacement of the midpoint greater than, equal to, or less than the amplitude of either original pulse? Do you think this displacement is greater than, equal to, or less than the difference between the amplitudes of the two original pulses?

13. Do the pulses that emerge from the overlap region look similar to those that enter?

DISCUSSION OF EXPERIMENT C-1

You saw in this experiment that a pulse moves along the spring without changing its shape appreciably. The amplitude decreases, but the decrease is not very big during one trip down the spring. Therefore, you can follow the pulse during a few trips back and forth. The time it takes for a complete round trip of a pulse is nearly constant; thus the velocity of the pulse does not depend much on the amplitude or the breadth of the pulse.

After a positive pulse reflects from one end, it comes back as a negative pulse and, similarly, a negative pulse is reflected as a positive pulse. That is, the direction of the displacement of the pulse is reversed upon reflection. However, the reflected pulse has the same general shape as the original pulse.

When two positive pulses overlap, they combine to produce a displacement that is larger than the amplitude of either original pulse. When one positive and one negative pulse overlap, they combine so that they almost cancel each other. It is consistent with your observations to say that at each point along the spring, the actual displacement is the sum of the individual displacements which the two pulses would cause separately at that moment. This statement is called the law of *superposition* for displacements. The pulses emerge from the region of overlap largely unchanged in size and shape.

Finally, you probably observed that the time it takes for a pulse to make one complete round trip along the spring is the same as the time for one complete oscillation of the vibration pattern corresponding to the first harmonic. All these observations suggest a model for the way simple patterns of vibration—called *standing waves*—result from the superposition of waves that travel back and forth along the spring. This same model applies to guitar strings.

Consider the wave train traveling to the right pictured in Figure 33a. For the time being, forget about the fact that the spring is only of length L and is tied down at the ends. Imagine for the moment that the spring is much longer.

Figure 33b shows how *another* wave train would appear if it were moving in a direction opposite to that shown in Figure 33a. Now imagine that these two waves are present simultaneously in the region between points B and C. This is the region where the real spring exists. Applying the principle of superposition, we conclude that the spring should appear at that instant as shown in Figure 33c. The displacements at both ends (points B and C, where the spring is tied down) are zero, and in the middle the displacement is twice that of either wave by itself.

A wave train like those shown in Figure 33 is a set of identical wave shapes, or *cycles*, which repeat themselves along the wave train. The length of one of these identical wave shapes is called the *wavelength* of the wave. As you can see in Figure 34, one wavelength, represented by the symbol λ (*lambda*), is the distance from crest to crest. The wavelength is also the distance from any point on one cycle of the wave to the corresponding* point on the next cycle.

Question 21. Using the definition of wavelength, explain why the distance from A to C in Figure 33a is one wavelength, and why the distance from D to C is one-half wavelength.

When the crests (or troughs) of two waves occur at the same place at a given time, as in Figure 33, we say that the waves are *in phase*.

Now consider how the two waves traveling in opposite directions would appear after each wave had traveled one-quarter of a wavelength farther than shown in Figure 33. Point A in Figure 35a shows the wave moving to the right after it has moved ¼ λ further to the right. In Figure 35b, the wave is moving to the left after it has moved ¼ λ further to the left.

We again imagine that these two waves

---

*Given a point on the wave, a *corresponding* point on the next cycle would be the next point on the wave where the slope has the same value and the amplitude has the same value.

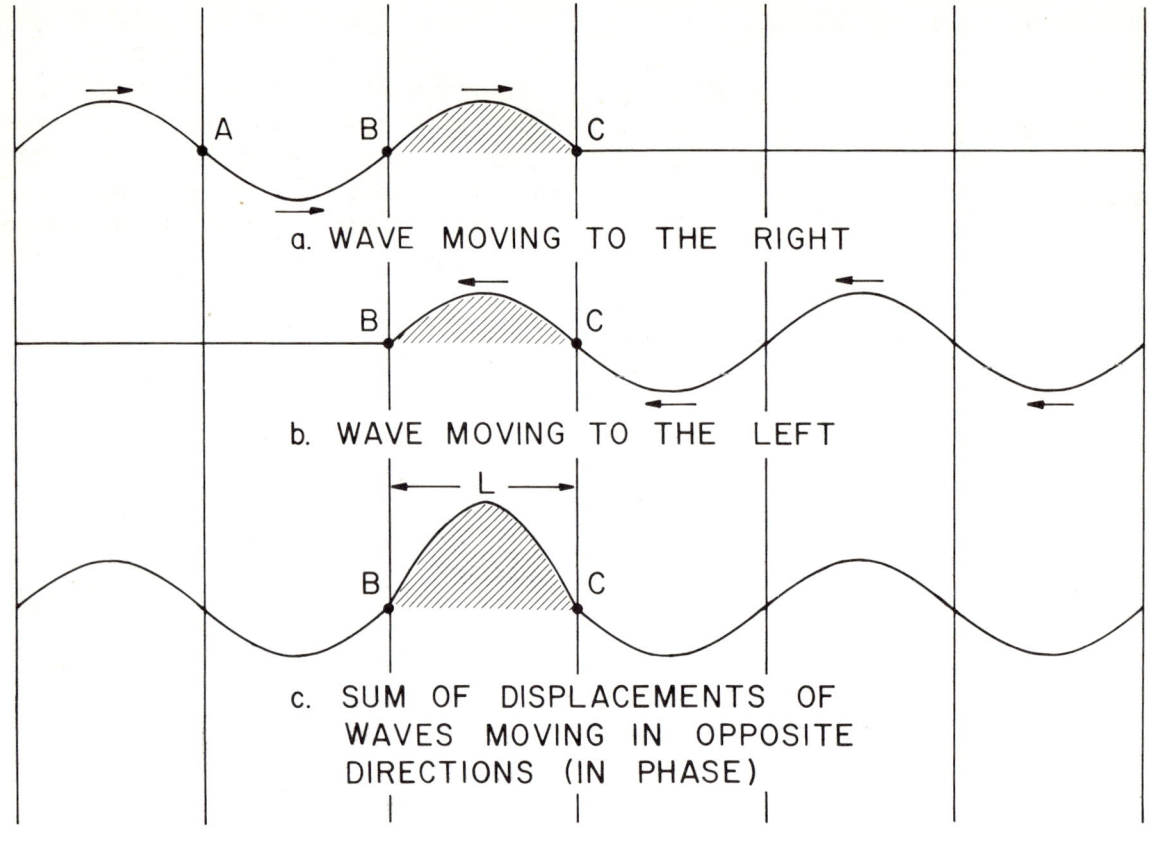

Figure 33.

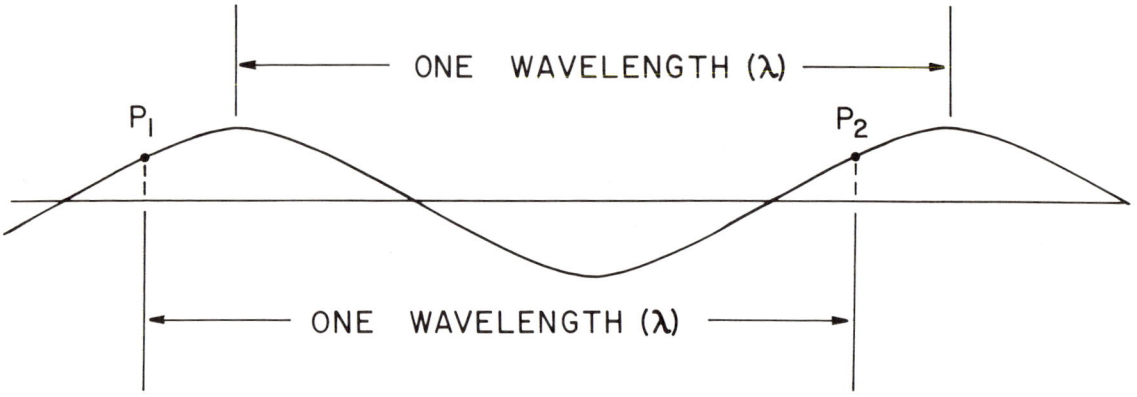

Figure 34.

are present simultaneously in the regions between points B and C, where the real spring exists. When we apply the principle of superposition to these two waves, the spring appears at that instant as shown in Figure 35c. One wave cancels out the other. We say that these two waves are out of phase by one-half wavelength.

Let us see what happens to the two waves after they have each traveled another quarter of a wavelength. Their positions would then appear as shown in Figure 36a and b. The two waves are again *in* phase, but each has displacement opposite to that it had initially. As shown in Figure 36c, applying the principle of superposition results in a trough with twice the amplitude of either wave separately.

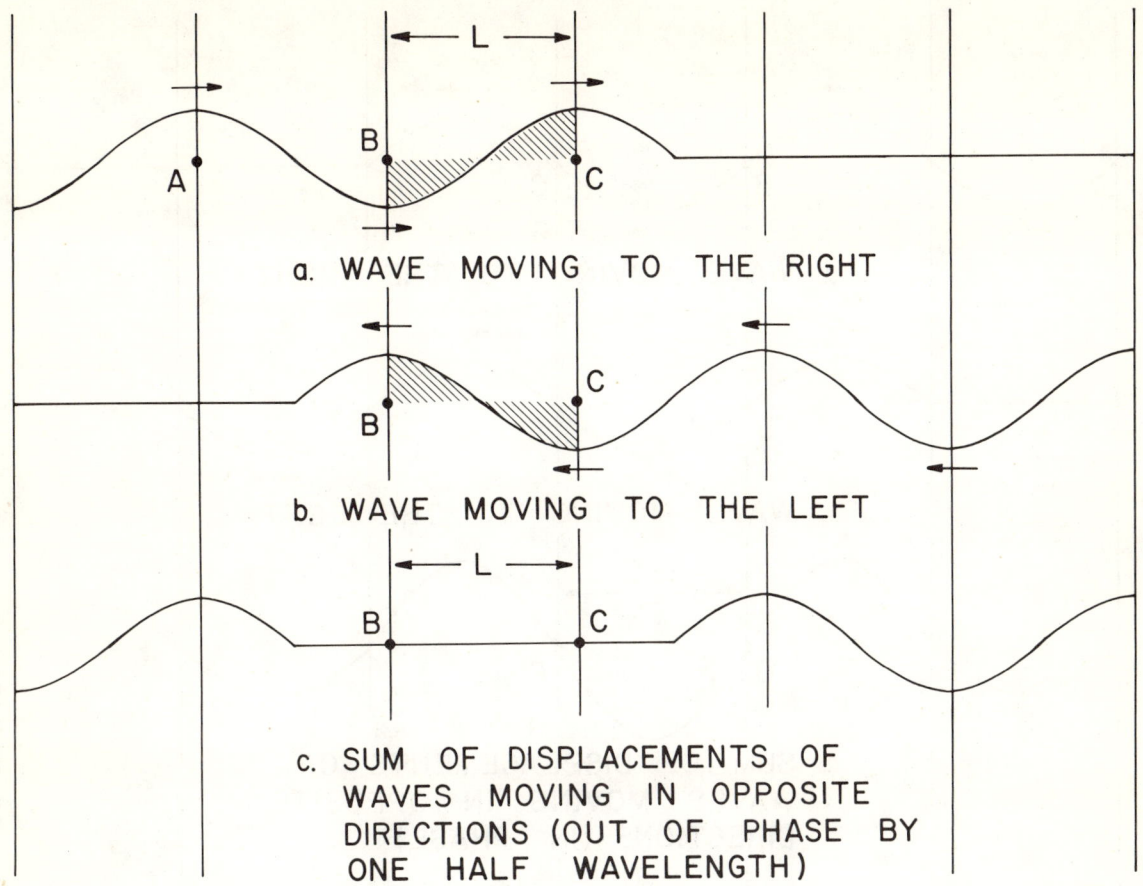

Figure 35.

Problem 13. Sketch the two waves, one moving to the right, and the other to the left a. after each wave has moved ¼ λ farther than shown in Figure 36, and b. after each wave has moved ½ λ farther than shown in Figure 36.

Question 22. Study the superposition that results between points B and C for Figures 33, 35, 36, and the two sketches you made in Problem 13. Regard the region between B and C as the location of the real spring. What is that part of the spring between B and C doing during this sequence? Are points B and C nodes, as the end points of the spring must be, during the time that the two imaginary waves are moving through each other?

Two waves moving in opposite directions on a long spring can, by the principle of superposition, be added algebraically. The superpositions described above reveal that two wave trains traveling in opposite directions can combine to produce a pattern on a coiled spring or a guitar string that includes nodes and antinodes at specific locations along the spring or string, just as we observed them on an actual vibrating spring. These patterns of vibration are referred to as *standing waves*. The two original moving wave trains are called *traveling waves*.

What are the implications of our inference that two oppositely directed but otherwise identical traveling waves are equivalent to a standing wave? If this inference is to be plausible, then it must be that the wave traveling in one direction is the excited wave. The wave traveling in the other direction can then only be the result of reflecting from the end of the spring or guitar string. Is this the case? Look at Figures 33a and 33b. The wave traveling to the right (shown in Figure 33a) if it had not encountered the tied down end at C, would have produced a negative displace-

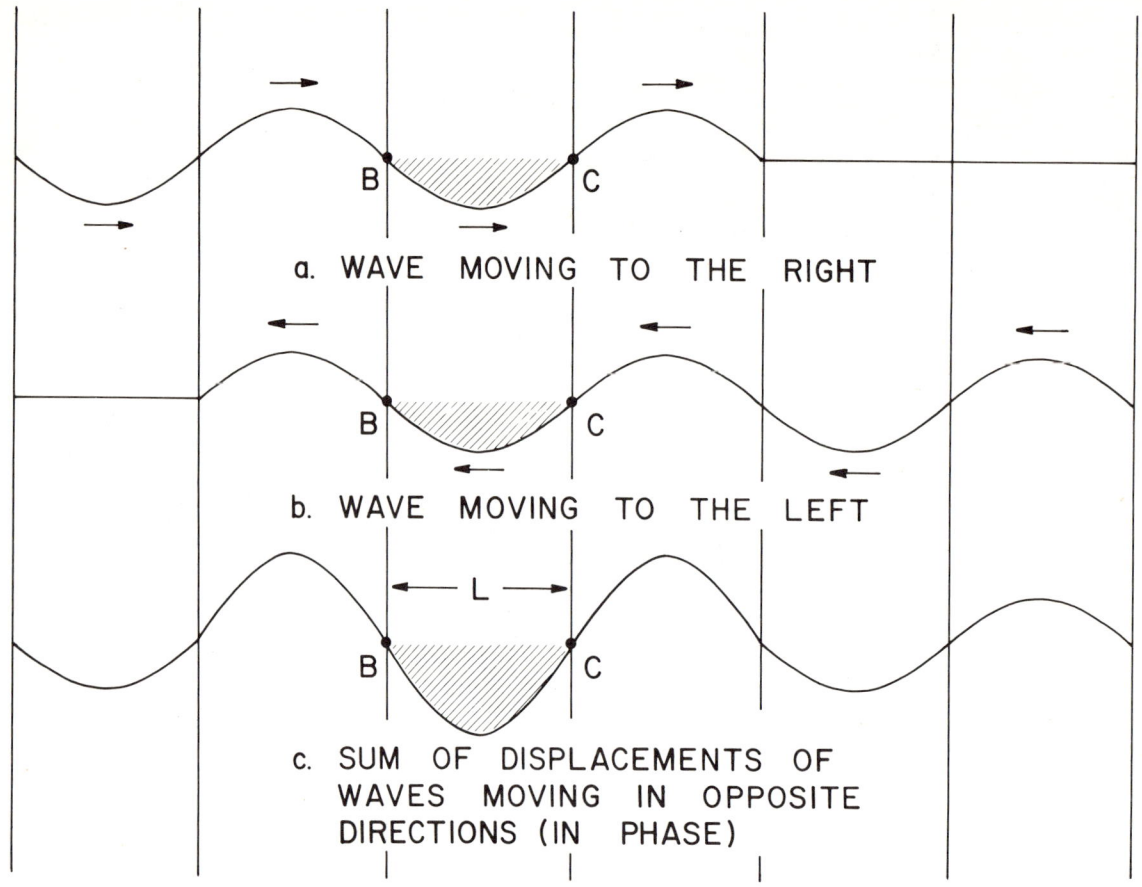

Figure 36.

ment a half wavelength to the right of C. For a reflection, we know that such a pulse has its displacement reversed at C and is sent back to the left. This positive pulse traveling to the left is exactly what is drawn in Figure 32b. As a result of reflections at B and C, the pulses shown in Figures 33a and 33b exist simultaneously in the region between B and C.

You found in Experiment C-1 that the time it takes for a pulse to make one complete round trip along a spring (or guitar string) is the same as the time for one complete oscillation of the vibration pattern corresponding to the first harmonic. You have also just seen that with our model of two waves traveling through each other to form standing waves, each wave travels one wavelength during the time that the standing wave makes one complete oscillation. Thus, the period of the oscillation of the spring is the same as the time it takes for one of these waves to move a distance of one wavelength along the spring.

The model we have created to account for the fundamental vibration pattern of a spring (or guitar string) gives results which are in agreement with experiments you have conducted only if the wavelength of the waves traveling back and forth on the spring is exactly twice the distance between the fixed ends of the spring:

$$\lambda = 2L \qquad (19)$$
(1st Harmonic)

According to our model, a guitar string vibration is in its *fundamental mode* whenever we have produced waves for which the wavelength is given by

$$\lambda = 2L$$

where L is the distance between the fixed ends of the guitar string (the distance from the bridge to the nut).

From observations made in Experiment C-1, and from the model of two traveling waves, we conclude that a traveling wave moves along a guitar string a distance of one wavelength in a time of one period of oscillation of the string. Because distance traveled is related to speed and time by the equation

$$\text{Distance} = \text{speed} \times \text{time}$$

or

$$\text{Speed} = \text{distance/time}$$

we can write an equation for the speed (v) of a wave on the guitar string:

$$v = \lambda/\tau$$

where τ is the period of oscillation. This equation can be rewritten as

$$v = \lambda(1/\tau) \qquad (20)$$

Question 23. Suppose that a guitar string has a period of 0.01 s per oscillation. What is the value of the quantity $1/\tau$, and what does this quantity represent?

In Question 23, the quantity $1/\tau$ is what we have called the *frequency f* of oscillation. In general, the frequency is related to the period by the equation

$$f = 1/\tau \qquad (21)$$

Replacing $1/\tau$ in Equation (20) by f from Equation (21), we have the important relationship:

$$v = \lambda f \qquad (22)$$

Question 24. According to Equation (22), what happens to the wavelength if we keep the oscillation frequency fixed, but increase the speed of the wave in the string? What happens to the frequency of oscillation if we decrease the wave speed, but keep the wavelength unchanged? In a given medium (where the wave speed is fixed) what happens to the wavelength if we go to a higher frequency of oscillation?

For the first harmonic, we know that

$$\lambda = 2L$$

and from Equation (22),

$$\lambda = v/f$$

Setting the right sides of each of these equations for λ equal to each other, we have

$$2L = v/f$$

Multiplying both sides of this equation by f and dividing by $2L$, we get

$$f = \frac{v}{2L} \qquad (23)$$
(1st Harmonic)

Equation (23) is a theoretical equation, based upon our model, for the frequency of oscillation of a string of length L. The next step is to compare this theoretical equation with the equation determined empirically in Experiment B-1:

$$f = K\sqrt{\frac{T}{mL}} \qquad (24)$$

The equations don't look much alike, do they? However, the theoretical equation contains the wave velocity v, and until we discover how v depends on properties of the string such as T, m, and L, we cannot expect the two equations to be identical.

We have called attention to the fact that any displaced portion of the string returns to equilibrium more rapidly as the tension increases. You observed in an experiment that when the tension of a coiled spring was increased, the wave speed increased. Also the wave speed is lower for more massive springs than for lighter springs. (By "massive" we mean the mass per unit length.)

Your experience with waves on springs

and strings, along with the model we have developed, provides a *qualitative* relationship between tension, mass, length, and wave speed. It is possible to derive a *quantitative* relationship between these variables which is consistent with what you expect qualitatively, using Newton's laws of motion. We will not carry out the derivations here but will simply assert the result without proof.*

$$v = \sqrt{\frac{T}{m/L}} = \sqrt{\frac{TL}{m}} \quad (25)$$

Now eliminating v between equations 23 and 25, we obtain

$$f = \frac{1}{2}\sqrt{\frac{T}{mL}}$$

If we use the subscript 1 on f to designate the first harmonic, this equation becomes

$$f_1 = \frac{1}{2}\sqrt{\frac{T}{mL}} \quad (26)$$
(1st Harmonic)

The theoretical equation (26) and our experimental equation (24) are identical, if we put $K = 1/2$. Is this value consistent with your results of Experiment B-1?

It looks as though we may have a good theory, but there are still questions that need to be answered. The theory predicts the first harmonic standing-wave patterns correctly, but what about higher modes? Figure 37 shows traveling waves with one-half the wavelength of those shown in Figures 33, 35, and 36. Notice that we have possible nodes at points B and C, and another midway between these points.

Problem 14. Sketch the appearance of the two traveling waves of Figure 37 and their superposition as they would appear at the time when each traveling wave has moved ¼ λ farther. Sketch them again when they have moved another ¼ λ.

---

*For a derivation see, for example, Sears & Zemansky, *University Physics*, Addison-Wesley Publishing Co.

Question 25. From Figure 37 and the results of Problem 14, what can you conclude about the locations of nodes and antinodes from point B to point C?

The pattern of nodes and antinodes produced by these two traveling waves is identical to the second harmonic you saw on the coiled spring and on the guitar string. How fast would waves with this wavelength travel on the string? You showed in Experiment C-1 that the speed of a pulse does *not* depend on either the amplitude or the breadth of the pulse. Equation (25) thus holds regardless of the wavelength. However, for the waves shown in Figure 37, one whole wavelength fits into the length L. Taking this to be the pattern for the second harmonic, the condition for the second harmonic becomes

$$\lambda_2 = L$$

or, in terms of a half-wavelength:

$$2(\lambda_2/2) = L \quad (27)$$
(2nd Harmonic)

Since the relationship

$$f = \frac{v}{\lambda}$$

is still valid, replacing λ by L, and using the subscript 2 to designate the second harmonic,

$$f_2 = \frac{v}{L} \quad (28)$$
(2nd Harmonic)

Finally, replacing the speed v by its value given in Equation (25), we have

$$f_2 = \sqrt{\frac{T}{mL}} \quad (29)$$
(2nd Harmonic)

which can be written

$$f_2 = 2f_1 \quad (30)$$

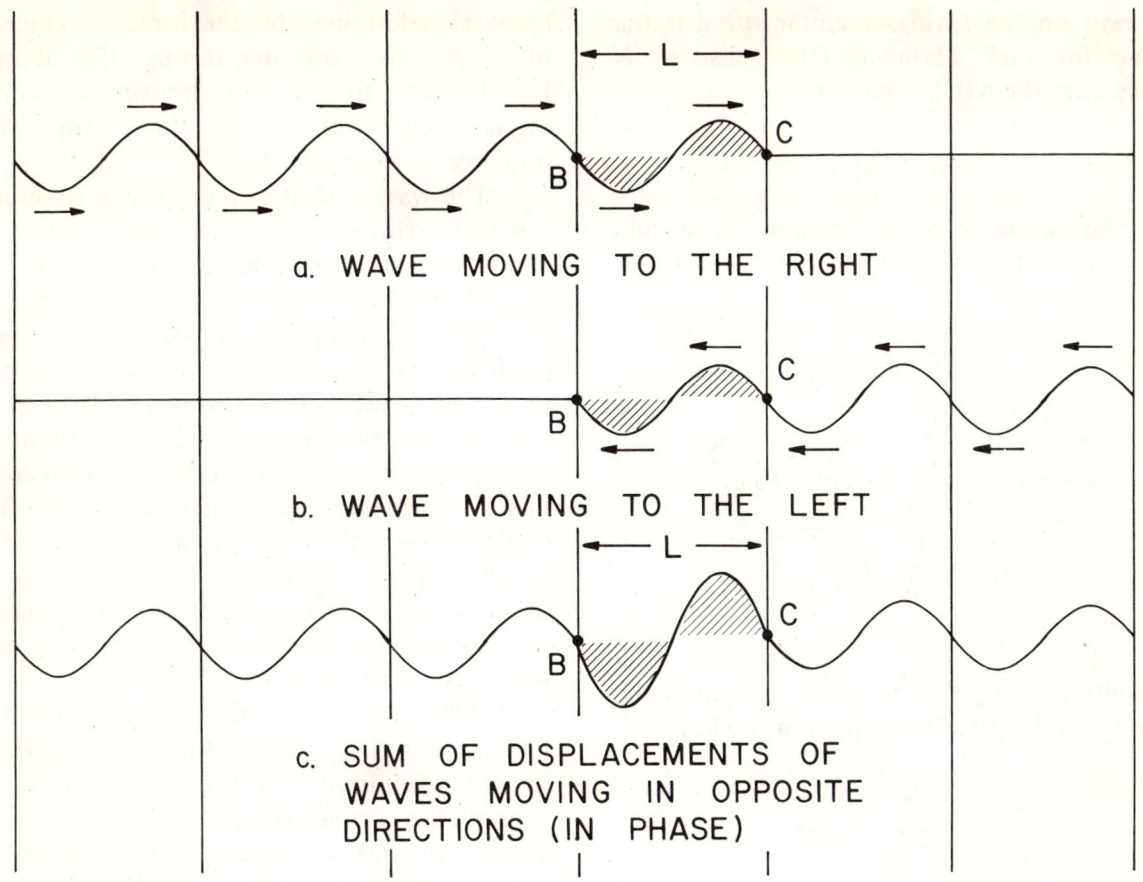

Figure 37.

These results agree with observations made during Experiment A-2: the frequency of the second harmonic is twice the frequency of the first harmonic.

If we consider the case of two traveling waves, each with one-third the wavelength of those we first discussed in Figure 32, we find that this situation produces a pattern identical to the third harmonic we observed in the spring, with three antinodes. The condition for the third harmonic standing wave is that three half-wavelengths fit exactly into the distance L between end points. As an equation, this condition is:

$$3 (\lambda_3/2) = L$$

Using the same method, we find the condition for the standing wave corresponding to the fourth harmonic to be

$$4 (\lambda_4/2) = L$$

Instead of writing a different equation for each condition, we may summarize these equations, and those for all the higher-order harmonics, into one equation. To do this we notice that the *order number* N of the harmonics (1st, 2nd, 3rd, etc.) is the number of half-wavelengths which must fit into the distance L between end points. As an equation, this condition becomes

$$N (\lambda_N/2) = L \tag{31}$$

Problem 15. For a certain guitar, it is 66 cm from the nut to the bridge along a guitar string. The fundamental frequency of this string is 220 Hz. What is the frequency of the string when there are nodes spaced 22 cm apart? What is the wavelength of the traveling waves on this string which give rise to this mode of oscillation?

You have already seen that the speed of

a wave on the spring or guitar string is the same for each harmonic (the subscript N designates the Nth harmonic),

$$v_N = f_N \lambda_N$$

In terms of string tension, mass, and length, the speed is given by

$$v_N = \sqrt{\frac{T}{m/L}}$$

Setting the right sides of these two equations equal, and solving for f_N, we have

$$f_N = \frac{1}{\lambda_N} \sqrt{\frac{T}{m/L}}$$

If we now solve Equation (31) for λ (resulting in $\lambda_N = 2L/N$) and use this value for λ_N in the preceding equation, we have

$$f_N = \frac{N}{2L} \sqrt{\frac{T}{m/L}}$$

or

$$f_N = \frac{N}{2} \sqrt{\frac{T}{mL}} \qquad (32)$$

Since the fundamental frequency is $f_1 = (1/2)\sqrt{T/mL}$, this can be written as:

$$f_N = N f_1 \qquad (33)$$

TRAVELING WAVES ON THE SOUND BOARD

Transverse pulses and waves travel along the sound board of a guitar, much as they do along a string. One complicating factor is that the sound board is a two-dimensional surface; therefore, waves can travel away from a source of disturbance in every direction in the plane of the sound board. When such traveling waves reach a boundary, such as the edge of the sound board, they are reflected, and the reflected waves superpose with the original waves. The resulting displacement at each point is determined by the forces produced by all the waves traveling through that point. If all these forces were known, the net displacement could be calculated from the principle of superposition.

The wave velocity of a traveling wave on a sound board depends on properties of the wood, primarily on its elasticity and its density. It also depends on the frequency. If we excite a disturbance by causing a local oscillation at a frequency f, the wavelength of the traveling waves that move away from the point of disturbance is given by the equation $\lambda = v/f$. If, in a given direction, it turns out that an integer number of half-wavelengths fit between two points that must be nodes (because they are clamped down or because they are too massive to move rapidly), then the principle of superposition predicts a standing wave pattern.

(When the original traveling wave and its reflection are in phase, producing a wave with twice the amplitude, the situation is known as *constructive interference*. When the two waves are out of phase, producing zero displacement, the situation is called *destructive interference*.)

If the half-wavelength is such that an integer number of them does not fit exactly between two points which must be nodes, the principle of superposition predicts that no standing-wave pattern is set up. In this case very little energy is transferred to the sound board. The two-dimensional vibrational patterns that can be excited at certain resonant frequencies on guitar sound boards are interesting. However, the solution of the two-dimensional problem is complicated.

MIXTURES OF HARMONICS

You have learned how standing waves can account for the various possible harmonics or modes of oscillation of a guitar string. However, we have discussed only situations where a single harmonic is present at a time. How can we pluck a guitar string in different positions, excite different sets of harmonics, and get the same pitch (that of the fundamental) but tones of different quality? Also, how does the pitch change suddenly when a vibrating string is touched at a certain point along its length?

It turns out that a plucked string gives several frequencies of sound at the same time: the fundamental and several harmonics. Does this fact mean that the string is vibrating with several frequencies at once? As a matter of fact, the string can vibrate with *any* number of the permissible standing-wave frequencies at the same time. Each possible standing wave is called a *mode of oscillation.* It is unusual for a plucked string to vibrate with just one of the simple wave shapes we have studied so far. A string plucked with a sharp point like a fingernail or guitar pick might, at different times, have the shapes shown in Figure 38.

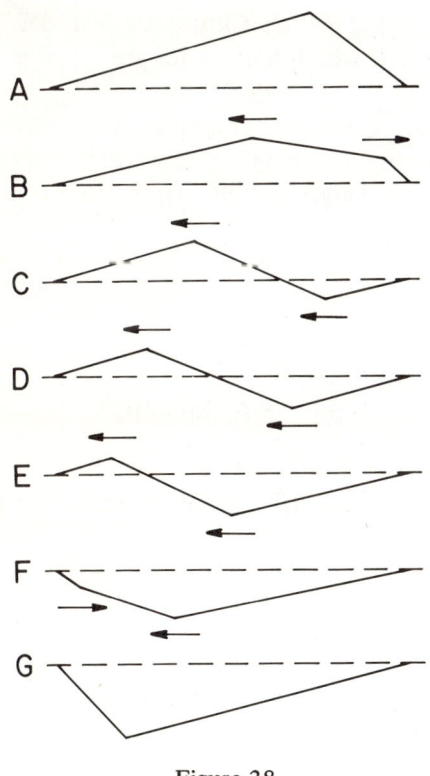

Figure 38.

At A, the string is at its initial displacement. At B, part of the disturbance has moved to the right, while the other part has moved to the left. In C, D, and E both disturbances are moving to the left. In F the disturbances are moving toward each other, and in G they are together. For *each* of these shapes of the guitar string, there is a set of harmonics (standing waves on the string), each with a different amplitude. The algebraic sum of these amplitudes gives the string shape at that instant. The nodes of each harmonic are in the usual positions on the string. But as the antinodes from the different harmonics vibrate at their various frequencies, the sum of all harmonics is the displacement shown in Figure 38, moving back and forth along the string. It is surprising that the sum of standing waves is a shape which is moving, but this is the case.

A simpler example will further illustrate this remarkable property of string oscillations. Figure 39 shows the shape of a guitar string plucked exactly at its midpoint by a sharp pick. The displacement is exaggerated to show the shape of the string.

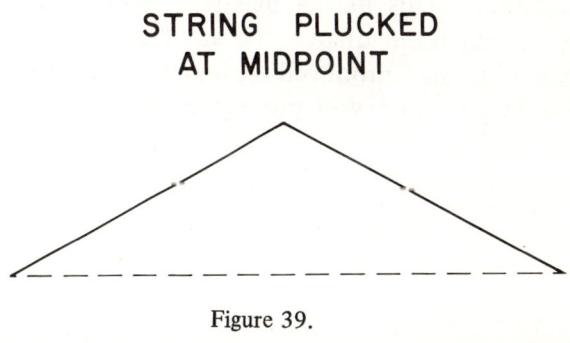

Figure 39.

The harmonics which add together to make this shape are shown in Figure 40. The amplitudes are drawn to scale.

Notice that the second, fourth, and sixth harmonics are missing. Higher odd-numbered harmonics, such as the ninth and eleventh, are present, but their amplitudes are small enough to ignore.

Problem 16. By measuring the displacements of the standing waves shown in Figure 40 from one fixed end to the other, sketch the algebraic sum of the first and third harmonics. How does the shape of this resulting wave compare with the shape of the guitar string shown in Figure 39?

Question 26. Why is the second harmonic absent when a guitar string is plucked in the middle by a sharp pick?

Problem 17. Try adding the fifth harmonic to the wave form you got in Problem 16. How does this change the wave form and how does the new shape compare with that of the string?

Figure 41 shows the relative amplitudes of harmonics which are produced when a string is plucked with a sharp pick one-quarter of the way from a fixed end. Again the sum of the amplitudes of the harmonics will give the original shape. But notice that the amplitudes of the second and third harmonics are greater than were the third and fifth harmonics when the string was plucked at its midpoint. This means that when you pluck the string at one-quarter the way from an end, the harmonics produced are louder than those produced when you pluck the string at the midpoint. This accounts for the tinniness of the sound when guitar strings are plucked near the bridge compared with the sound produced when the string is plucked at its midpoint. Electric guitars take advantage of this property by providing two or three pickups at different positions beneath the strings. At positions closer to the bridge, you can pick up more harmonics which are louder than at positions farther from the bridge.

The principle that a disturbance of any shape on a string can be represented by an algebraic sum of certain harmonics having certain relative amplitudes is called *Fourier's Theorem*. This theorem has wide application in acoustics, electronics, light, and anywhere else where wave motion is involved.

The point at which you pluck a guitar string *cannot* be a node because you have already displaced it. Thus if you know where a string has been plucked you know a point on the string which cannot be a node. Thus you can tell which harmonics are missing.

Example Question. Suppose you strike a piano string at a single point which is exactly 1/7 of the length of the string from one end

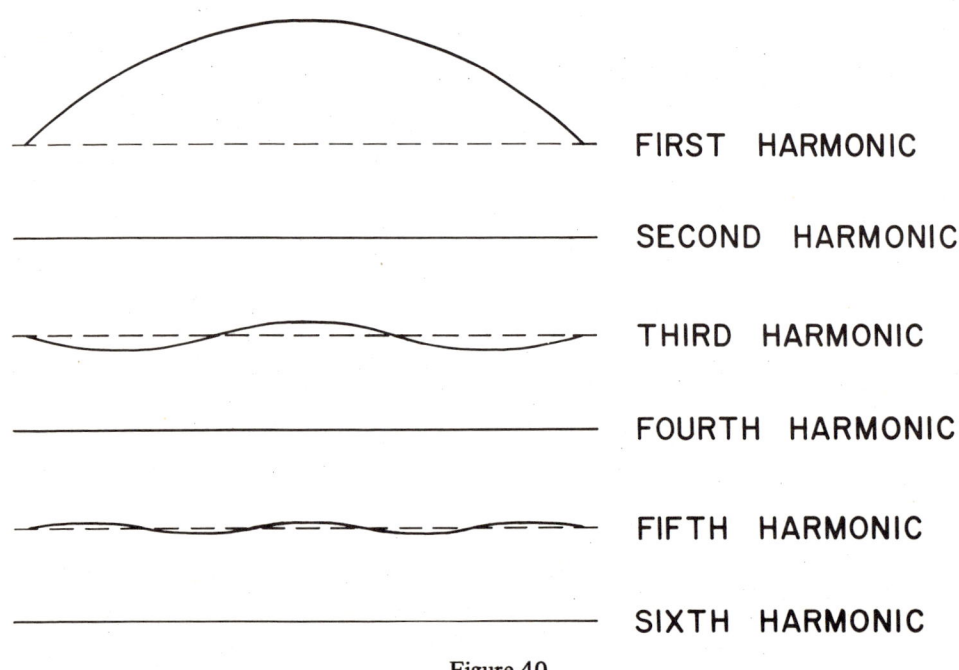

Figure 40.

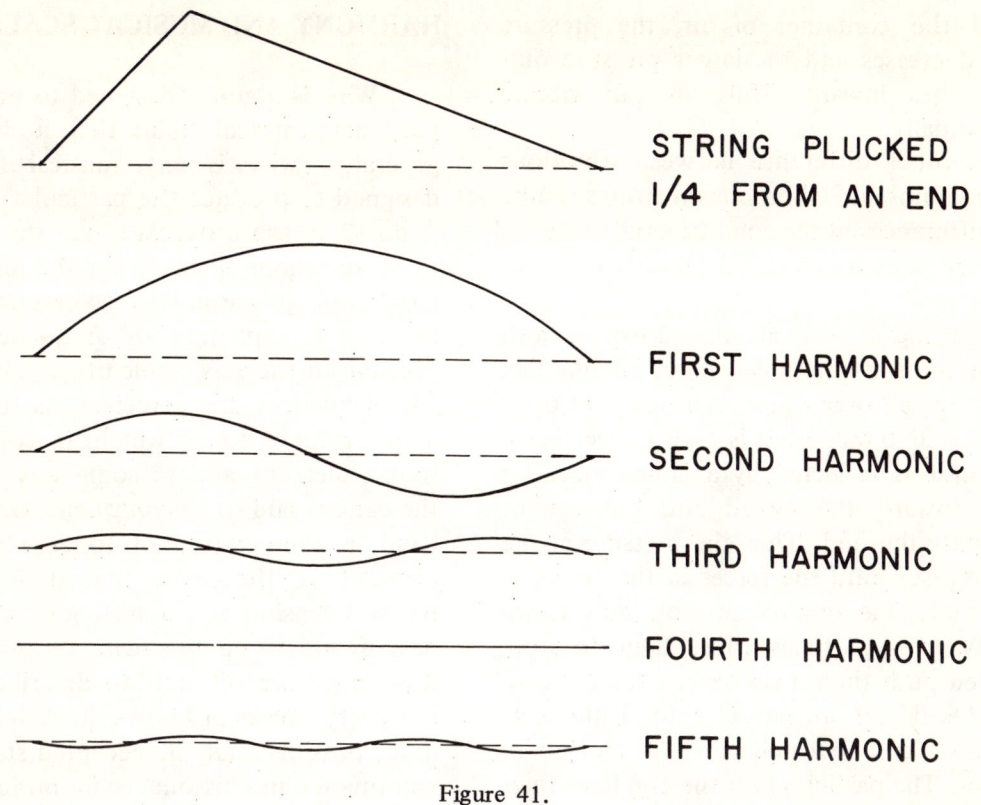

Figure 41.

of the string. What string modes would you expect to be missing from the vibration of this string?

At 1/7 of the way from one end of the piano string there cannot be a node. If there were a node at this position, there would be 7 antinodes between the fixed ends of the string. Thus the 7th harmonic would be present if a node existed 1/7 of the way from one end. Because the string is struck at this point, a node cannot exist there, and the 7th harmonic must be missing. Since the 14th, 21st, and all other multiples of 7 also would have a node at this position, these harmonics must also be missing.

Question 27. A guitar string is plucked 11 cm from the bridge. If the distance along the string from the bridge to the nut is 66 cm, what harmonics are missing? If the fundamental frequency of the string is 167 Hz (low E), what frequencies (called *overtones*) are missing from the harmonics of the string? What are some of the overtones which are present?

LONGITUDINAL SOUND WAVES IN AIR

Air has elasticity and mass, and thus it can conduct sound waves. If the air is confined, as in the sound box of a guitar, traveling waves can reflect and, by superposition, produce standing-wave resonances. However, there is an important difference between sound waves in air and the waves we have discussed thus far. Sound waves in air are *longitudinal,* not transverse. This means that the oscillations of air particles are in the same direction as the direction of travel of the wave. Air molecules that are displaced sideways do not experience restoring forces that pull them back, and thus no transverse oscillations occur in air. However, air *is* elastic. If you compress a container of air, the pressure on the air inside pushes outward; if you expand the container of air, the pressure inside decreases and the larger pressure outside pushes inward. Thus air can vibrate longitudinally.

(Another difference between vibrations in air columns and vibrations on strings results from differences in the conditions at the ends. These are sometimes called *boundary conditions*.)

A string is normally tied down at both ends, so the ends are nodes. A tube containing air, such as an organ pipe, can be either open or closed at the ends. It is easy to see that a wave pulse is reflected by a closed end. Air moves toward the closed end but cannot move past the end. Thus the pressure at the end increases until the forces in the direction opposite to the original motion slow down the advancing particles, cause them to stop, and then push them back away from the end. At a closed end, air particles don't move, so the closed end is a node for the motion of particles. The particles near the end have their direction of motion reversed, and the reflected pulse has a displacement opposite to that of the original pulse. If one thinks of a sound wave as an oscillation in pressure, instead of as an oscillation in the motion of particles (it is, of course, both), then a closed end is a place where pressure oscillations are maximum; that is, an antinode.

The velocity of sound in air at sea level is between 330 m/s and 340 m/s; it increases with temperature. At a fixed temperature, sound waves of all frequencies travel at the same speed. This speed is not too different from the speed of transverse waves in a guitar sound board. This means that the air in the sound box can have standing waves which resonate at frequencies that are in the same general range as the resonant frequencies of the sound board, since both vibrating structures have the same dimensions. Instrument designers over many years, have learned, by trial and error, to produce instruments for which the elasticity and thickness of the wood and the construction of the instrument are such that sound board, strings, and the air in the sound box have overlapping resonances.

HARMONY AND MUSICAL SCALES

Why is a guitar designed to produce the particular musical tones that it does? More generally, why is any musical instrument designed to produce the particular tones that it does? In the broadest sense, the answer to these questions depends on the answer to a more basic question. Why do certain combinations and sequences of frequencies sound pleasing to the ears, while other combinations and sequences are perceived as unpleasant? Two or more tones which, in combination, sound pleasant and in some way "final" to the ear are said to be *consonant*. On the other hand, a combination of tones which is unpleasant in the sense that it suggests unrelieved tension and a feeling of incompleteness is said to be *dissonant*. Consonance and dissonance are difficult to describe verbally, but every musician knows the difference. It is quite possible that the accepted standards of consonance and dissonance in music are in the process of changing.

Learning and experience play an important role in whether combinations of notes sound consonant or dissonant. However, for trained musicians raised in Western cultures, frequencies which are most consonant with each other are found to be in the ratio of small whole numbers. The smaller the numbers, the better the consonance; for example, tones whose fundamental frequencies are in the ratio of 3:2 are consonant. A *musical interval* is thus defined as the ratio of two frequencies, rather than the difference. This discovery of a connection between integer ratios and consonance was first made by the ancient Greeks, using tones from stretched strings. (They actually used ratios of string lengths, which they could measure, rather than frequencies, which they couldn't.)

Table III shows integer ratios (musical intervals) of frequencies, arranged in order from the integer ratio containing the smallest number to the integer ratio containing the largest number. The name shown next to each ratio is the musical name of the interval between the two notes. The intervals are arranged in order of increasing dissonance. That is, a minor sixth (8:5) sounds more dissonant than a minor third (6:5).

The actual frequencies used in music have varied with time and place, but once a standard tone is fixed in frequency, the ratios of Table III fix the frequencies of notes which are consonant with the first.

Table III. Musical Intervals

Name	Ratio of Frequencies
Unison	1:1
Octave	2:1
Fifth	3:2
Fourth	4:3
Major Second	5:4
Major Sixth	5:3
Minor Third	6:5
Minor Sixth	8:5
Second	9:8

Question 28. Two notes form an octave and a fifth, respectively, with a given note. What interval do these notes form with each other?

By a 1939 international agreement, the frequency of the note A above middle C is fixed at 440 Hz. This choice, along with the intervals, gives rise to the *diatonic scale* shown in Table IV. Each frequency of the scale is consonant either with middle C or with a note consonant with middle C. For example G is consonant with C (ratio of 3:2), but D is not. However, G and D are consonant (ratio of 4:3).

Problem 18. Find the interval which each note in the diatonic scale forms with middle C. Which of these are most consonant? Which are least consonant?

Many theories have tried to explain why tones whose fundamental frequencies are in

Table IV. The Diatonic Scale

Note	Frequency
C	264 Hz
D	297
E	330
F	352
G	396
A	440
B	495
C'	528

the ratio of small whole numbers sound "right" together while others don't. The theory based on *beat frequencies* between overtones, due to Hermann L. F. von Helmholtz (1821–1894), has been fairly successful, but it probably is not strictly correct.

Beat frequencies are the low frequencies that you hear as you bring two frequencies closer and closer together. The *beat frequency* between two frequencies, f_1 and f_2, is their difference,

$$\Delta f = f_2 - f_1$$

For example, the beat frequency between a 340-Hz tone and a 320-Hz tone is 20 Hz. When comparing two tones, one says that the tones have the same pitch if the beat frequency is very small or zero. (You can hear a beat frequency by doing the following: slightly detune the sixth string of a guitar so that the pitch produced while pressing it at the fifth fret is slightly different than the open-string pitch of the fifth string (A). Then, holding down the sixth string at the fifth fret and leaving the fifth string open, strongly pluck both strings simultaneously. The alternately loud and soft sound you hear is the beat frequency. Experiment with changing the beat frequency.)

In Helmholtz' view, dissonance is due to rapid beats between the overtones of musical sounds. When two tones have fundamental frequencies in the ratio of small integers, many of their harmonics coincide. They do not produce so many distracting beats and do not sound dissonant. For example, consider D and C of the diatonic scale, which are in the ratio 9:8. When these notes are sounded on two different guitar strings, the fundamentals differ by 33 Hz, which is in the harshest region of beat frequencies. D is in the scale, not because of any consonance with C, but rather, as we have already pointed out, because G forms a perfect fourth (ratio of 4:3) with D, and G in turn forms a perfect fifth (ratio of 3:2) with C.

A more pertinent example might be two strings with fundamental frequencies of 380 Hz and 264 Hz. These frequencies are in the ratio of 95:66. A ratio of 99:66 would be a fifth. Why do these two strings sound dissonant? The fundamentals are far enough apart so that they don't produce a noticeable beat. However, the second harmonic of 380 Hz (2 × 380 = 760 Hz) and the third harmonic of 264 Hz (3 × 264 = 792 Hz) produce a beat frequency of 792 − 760 = 32 Hz, which does produce a sensation of dissonance. In contrast, every second harmonic of G (which forms a perfect fifth with C) coincides exactly with every third harmonic of C, as indicated in Table V. When the harmonics do differ, it is always by a multiple of 132. Thus there are no harmonics which give beats at a frequency of less than 132 Hz, and this does not sound dissonant.

(You should realize that, as we have already hinted, consonance and dissonance, as well as the structure of musical scales, are strongly influenced by culture. These intervals and scales have evolved at least partially because of what has been considered in Western culture to be music. In Oriental cultures, very different [more complex] scales have developed, and different ideas of dissonance and consonance are held.)

A guitar tuned to produce the frequencies on the diatonic scale is said to be tuned in *just intonation*. The open strings of a guitar tuned to the diatonic scale would have notes corresponding to frequencies listed in Table VI (notes in octaves below middle C are indicated by multiple letters, while those in octaves above are written with primes). These notes are in the diatonic scale of C, but they

Table V. Comparison of the Harmonics of C and G

C	G
f_1 = 264 Hz	f_1' = 396 Hz
f_2 = 528 Hz	
f_3 = 792 Hz	f_2' = 792 Hz
f_4 = 1056 Hz	f_3' = 1188 Hz
f_5 = 1320 Hz	
f_6 = 1584 Hz	f_4' = 1584 Hz

Table VI. Open Guitar String Frequencies

Note	Frequencies (Hz)
EE	165
AA	220
D	297
G	396
B	495
E′	660

are spread through two octaves in order to give an appropriate range. The other notes are obtained by stopping the strings at frets.

When notes whose frequencies are related to one another by certain simple ratios are played at the same time on a piano, they sound harmonious. One set of three notes is called the *tonic chord of the key of C major*. It consists of three white piano keys, C, E, and G. These notes have fundamental frequencies of 264 Hz, 330 Hz, and 396 Hz, which are in the proportion 4:5:6. On a piano, two keys that are one octave apart are separated by eleven other keys. If the piano were tuned to produce perfect harmonies in certain chords, such as those for the key of C major, the frequency ratio between two adjacent keys would vary from pair to pair along the keyboard. This would make the piano sound fine for songs written so as to emphasize those particular chords (for example, in the key of C and F), but not so harmonious if other chord structures are emphasized (as in a song written in the key of D sharp).

Another problem is that pianos frequently accompany other instruments, and if instruments using some other system are to sound harmonious when played together with the piano, some compromise must be made. For example, a violin can play the notes for any chord, since the length of string vibrating can be varied continuously. However, a piano has only a fixed number of notes, as does a stringed instrument with frets, like a guitar. The simplest compromise, and one that has been adopted by virtually all musicians in the Western world, is to make the frequency ratio between every two adjacent notes the same. A scale constructed this way is called an *equally tempered* scale. Since there are twelve different notes per octave, there are twelve ratios between the notes in one octave. If these ratios are all equal to each other, as required for an even-tempered scale, we can represent the constant ratio with the symbol R. Knowledge of the frequency of the beginning note in an octave allows us to find the frequency of the next note by multiplying by R. Thus if f_0 is the frequency of the beginning note, $f_0 \cdot R$ is the frequency of the first note above the beginning note of the octave. The frequency of the second note above is found by multiplying the frequency of the first note by the ratio R. Thus the second note frequency is

$$f_0 R \times R = f_0 R^2$$

In a similar way, the frequency of the third note above the beginning note is

$$f_0 R^2 \times R = f_0 R^3$$

By repeated calculations like these, we find that the 12th note above the beginning note has a frequency given by

$$f_0 R^{12}$$

If we now form the ratio of the frequency of the 12th note to that of the beginning note, we have

$$f_0 R^{12} / f_0$$

But this ratio reduces to just R^{12}. You already know that the frequency ratio of the 12th note to that of the beginning note must be 2 to 1 (which has a value of 2), since the notes are an octave apart. Therefore

$$R^{12} = 2$$

This equation can be solved by taking the 12th root of both sides, with the result that

$$R = \sqrt[12]{2}$$

A nine-place electronic calculator gives, for $\sqrt[12]{2}$, the value 1.059463094. Thus, as a sufficiently precise approximation,

$$R = 1.05946$$

To build a scale using this ratio, we must agree on the frequency of *one* note. By agreement that note is called *concert* A, and is always set at 440 Hz. Table VII shows the frequencies of the thirteen notes from middle C to C′ an octave higher. The frequencies of notes in the equally tempered octave are slightly lower than the frequencies of the corresponding notes on the diatonic scale described earlier, but the differences are less than 1% and cannot be detected by most humans. Note also that the frequency ratio between G and C, for example, which was 3 to 2 or 1.500 for the diatonic scale, comes out to be 1.498 for the equally tempered scale. Ratios like this one are close enough to ratios of small integers that their notes still sound consonant.

Problem 19. Calculate the ratio between frequencies of E and F and between those of B and C′ on the diatonic scale shown in Table IV. How do these ratios compare? Why can't this ratio be used between each of the twelve notes in an octave on the piano?

In a symphony orchestra, most musicians tune their instruments so that they conform to the oboe. You have undoubtedly heard this take place while an orchestra is warming up. For stringed instruments, all that is required is that the fundamental of each

Table VII.

Key Symbol	Frequency
C (white key)	261.6
C# (black key)	277.2
D (white key)	293.7
D# (black key)	311.1
E (white key)	329.6
F (white key)	349.2
F# (black key)	370.0
G (white key)	392.0
G# (black key)	415.3
A (white key)	440.0
A# (black key)	466.2
B (white key)	493.9
C (white key)	523.3

open (unshortened) string should have the same pitch as the corresponding note on the oboe. Experienced musicians then know where to put their fingers in order to produce any other desired note. However, the frets on a guitar provide a fixed number of places (frets) to put your fingers, which constitute a scale. It is not necessarily true, however, that the notes of this scale correspond in a one-to-one way with the notes on a piano. The guitar builder may have chosen a different scheme, or perhaps simply failed to achieve fret positions that would produce the intended sequence. If the guitar were tuned so that each open string note had the same frequency as the corresponding key on a piano tuned to the equally tempered scale, the open string guitar frequencies would be as follows:

Note	Frequencies (Hz)
EE	164.8
AA	220.0
D	293.7
G	392.0
B	493.9
E′	659.3

In Experiment C-2, you will have an opportunity to find out for yourself what scale has been used on the guitar.

EXPERIMENT C-2. Guitar Scales

All you need to do this experiment is a guitar and a good meter stick. Measure the distance between the bridge and each fret. Number the frets, starting with 1 for the fret nearest the nut. Fill in the blanks on Table VIII. The column of distances represents the data from this experiment. Assume that the A string of the guitar is tuned to 220 Hz. This is the frequency of the fundamental of the tone made by this string when it is not touched, or stopped, anywhere. Other entries in this column can be calculated from Equation (3):

$$f = \frac{C}{L}$$

Since $f_0 = C/L_0$, where L_0 is the full length of the string, $f_N = (L_0/L_N) f_0$, where N = 1, 2, 3, ...

Now answer these questions:

1. Is the scale on your guitar an equally tempered scale? That is, are all the ratios in the last column equal or nearly equal?

2. Are any or all of these ratios equal or nearly equal to the ratio between the frequencies of two adjacent piano keys (1.0595)?

3. Could you play the note A one octave higher by stopping the A string at a fret? Which fret? At what distance along the string is this fret located?

4. How many notes are there to an octave? That is, how many frets are there between the nut and the midpoint, counting the one at the midpoint?

5. Is there a fret at the node closest to the nut for the third-harmonic vibration? For the fourth-harmonic vibration? For the ninth-harmonic vibration? If so, which frets are these?

6. Based on your data, would you predict that the tones made in the following ways are one octave apart?

    a. Hold the A string down at fret 1 and strum the string at its midpoint.

    b. Hold the A string at fret 13, and strum the midpoint of the remaining string (toward the bridge).

Test your prediction by holding the A string down at the two frets described in Question 6 and comparing the sounds.

Table VIII.

Fret No.	Distance from Fret to Bridge (in cm)	Frequency of a String Stopped at Fret (in Hz)	Ratio of This Frequency to Previous Frequency (Round to 3 Figures)
0 (open)		220 (f_0)	
1			
2			
3			
4			
5			
6			
7			
8			
9			
10			
11			
12			
13			
14			
15			
16			
17			
18			
19			
20			

SUMMARY

The six different open strings of the guitar produce six different pitches. The thicker strings produce lower frequencies because their mass per unit length is greater. To tune the guitar to the proper frequency the tension of the strings is changed by using the tuning pegs. Increasing the tension increases the pitch. The relative proportions of the sound-wave harmonics are determined by the material of the strings, the natural frequencies of the sound board and cavity, and the method of plucking. These variables account for the differences in sound quality obtainable.

When playing the guitar, the effective lengths of the strings are altered by holding them down against frets, which changes their fundamental frequencies.

In this section you have learned a theory that two waves traveling in opposite directions can superimpose to produce standing waves. The condition required for standing waves on strings fixed at both ends is

$$N(\lambda_N/2) = L$$

where N is an integer, λ_N is wavelength, and L is the length of the string. This equation says that if you can fit N half-wavelengths in a length L, you will have standing waves.

Using this theory, we could derive an equation which accounts for the string equation determined empirically in Section B:

$$f = K\sqrt{\frac{T}{mL}}$$

You also learned that mixtures of harmonics are present on a vibrating string because these harmonics must add up to give the shape of the string at any given time. This effect accounts for the different amplitudes of harmonics for different positions at which a string is plucked, and for the different ways it is plucked (with thumb or pick).

Each mode of vibration receives a certain amount of energy when a string is plucked, and each mode has its own life history, independent of the others. Some string modes die out more quickly than others, and some transmit their energy more readily to the air through the bridge and guitar body, but no energy is exchanged between modes. Each string may be thought of as a number of independent oscillating systems, each with its own separate single frequency.

Finally, you learned that harmony was related to *ratios* of fundamental frequencies of tones, and that musical scales could be constructed to produce harmonics. You found that the guitar is constructed to produce an equally tempered scale the same as a piano tuned to that note.

Work Sheet
EXPERIMENT A-1

Name _____

1. _____
   _____

2. _____
   _____
   _____
   _____

3. _____
   _____

4. _____
   _____

5. _____
   _____
   _____
   _____

6. _____
   _____
   _____
   _____

7. _____
   _____
   _____
   _____

8. _____
   _____
   _____

9. _____
   _____
   _____
   _____
   _____

10. _____
    _____
    _____

11. _____
    _____

12. _____
    _____

13. _____
    _____
    _____
    _____

14. _____   16. _____
    _____   _____
15. _____   _____
    _____   _____
    _____   _____

Work Sheet
EXPERIMENT A-2

Name _____

1.

2. _____
   _____
   _____

3. _____
4. _____
   _____
5. _____
   _____
   _____

7.

8. _____
   _____
9. _____
10. _____
    _____
11. _____
12. _____
    _____
    _____

13. _____
14.

15. _____
    _____
16. _____
    _____
17. _____
18. _____
19.

20. _____
    _____
21. _____
    _____
22. _____
    _____
23.

24. _____
    _____
25. _____
    _____
    _____

26. _____   28. _____
    _____   _____

27. _____   _____
    _____

Work Sheet
EXPERIMENT B-1

Name _____

1–10.

Table 1.

$T =$ _____ N $L =$ _____ m

String	Frequency (Hz)	Total Mass (kg)	Total Length (m)	$\dfrac{\text{Mass (kg)}}{\text{Length (m)}}$	Vibrating Mass (kg)
low E					
A					
D					
G					
B					
high E					

11–13.

Table 2.

$m =$ _____ kg $L =$ _____ m

$m/L =$ _____ kg/m

	Frequency (Hz)	Tension (N)
1 kg		
2 kg		
3 kg		
4 kg		
5 kg		
6 kg		

14–15.

Table 3.

$T =$ _____ N $m/L =$ _____ kg/m

Frequency (Hz)	Length (m)

16. _____
17. _____
18. _____
19. _____
20. _____
   _____
   _____
21. _____

22. _____
23. _____
24. _____
   _____
25. _____
26. _____
27. _____

Work Sheet
EXPERIMENT B-2

Name _____

Part I

1. _____
   _____

2. _____
   _____

3. _____
   _____
   _____

4. _____

5. _____

6. _____

7. _____
   _____
   _____

8. _____

9. _____

10. _____
    _____
    _____

11. _____
    _____
    _____

12. _____
    _____
    _____

Part II

1. _____

2. f_{EE} (Hz) = _____
   f_{AA} (Hz) = _____
   f_G (Hz) = _____
   f_B (Hz) = _____
   $f_{E'}$ (Hz) = _____

3.

4. | Frequency (Hz) | Meter Reading (dB) |
   |---|---|
   | | |
   | | |
   | | |
   | | |
   | | |
   | | |
   | | |
   | | |
   | | |
   | | |

5. _____

6. _____
   _____

7. _____

8. _____
   _____
   _____

COMPUTATION SHEET

Work Sheet
EXPERIMENT C-1

Name _____

1. _____
2. _____
3. _____
4. _____
   _____
5. _____
   _____
   _____
   _____
6. _____
   _____
   _____
7. _____
   _____
   _____
   _____

8. _____
   _____
9. _____
   _____
10. _____
    _____
    _____
    _____
11. _____
12. _____
    _____
    _____
13. _____
    _____
    _____

COMPUTATION SHEET

Work Sheet
EXPERIMENT C-2

Name _____

1. _____
   _____

2. _____
   _____

3. _____
   _____

4. _____
   _____

5. _____
   _____
   _____
   _____

6. _____
   _____
   _____
   _____

T5-AGE-485

0-07-001716-6